Pathogen removal using saturated sand columns

supplemented with hydrochar

Jae Wook Chung

Thesis committee

Promotor

Prof. Dr P.N.L. Lens
Professor of Environmental Biotechnology
UNESCO-IHE, Institute for Water Education, Delft

Co-promotor

Dr J.W. Foppen
Associate Professor of Hydrology
UNESCO-IHE, Institute for Water Education, Delft

Other members

Prof.Dr T. W. Kuyper, Wageningen University
Prof.Dr H. Harms, Helmholtz Centre for Environmental Research, Leipzig, Germany
Prof. Dr J.F. Schijven, Utrecht University
Dr S. Rutjes, National Institute for Public Health and the Environment, Bilthoven

This research was conducted under the auspices of the Graduate School for Socio-Economic and Natural Sciences of the Environment (SENSE)

Pathogen removal using saturated sand columns supplemented with hydrochar

Thesis

submitted in fulfilment of the requirements of

the Academic Board of Wageningen University and

the Academic Board of the UNESCO-IHE Institute for Water Education

for the degree of doctor

to be defended in public

on Friday, 30 October 2015 at 01:30 p.m.

in Delft, the Netherlands

by

Jae Wook Chung,

Born in Seoul, Republic of Korea

CRC Press
Taylor & Francis Group
Boca Raton London New York

CRC Press is an imprint of the
Taylor & Francis Group, an **informa** business
A BALKEMA BOOK

First issued in hardback 2018

CRC Press/Balkema is an imprint of the Taylor & Francis Group, an informa business

© 2015, Jae Wook Chung

Published by:
CRC Press/Balkema
PO BOX 11320, 2301 EH Leiden, The Netherlands
e-mail: Pub.NL@taylorandfrancis.com
www.crcpress.com-www.taylorandfrancis.com

ISBN 13: 978-1-138-38169-8 (hbk)
ISBN 13: 978-1-138-02929-3 (pbk)

Acknowledgements

First and foremost, I wish to express my special appreciation to promoter Prof. Dr. ir. Piet Lens and supervisor Assoc. Prof. Dr. Jan Willem Foppen for scientific guidance and encouragement throughout my PhD period. I would never have been able to finish this research without their contribution. Also, their enthusiasm and a good sense of humour have vitalized my staying in Delft.

I would like to acknowledge Dr. Saskia Rutjes and Prof. Dr. Jack Schijven for their academic advices and provision of virus stocks. I am also grateful to Fred Kruis, Peter Heerings, Lyzette Robbemont, Frank Wiegman, Ferdi Battes and Berend Lolkema of UNESCO-IHE for their assistance in laboratory experiments. Also, I thank Jolanda Boots for her administrative works during my PhD period.

Many thanks to my colleagues Lucian Staicu, Heyddy Calderon, Angela Ortigara, Pimluck Kijjanapanich, Suthee Janyasuthiwong, Joana Cassidy, Rohan Jain, Maribel Pereyra, Angélica Rada, Erika Espinosa and SangYeob Kim with whom I had been through both good and bad days. The time we spent together in Delft will be missed.

I gratefully acknowledge MangU Jeja Church. Korean Church of Brussels and Korean Reformed Church of Rotterdam for accompanying me throughout this journey.

Lastly, I would like to thank my dearest family for their love, prayers, encouragement and patience.

Abstract

Sufficient provision of clean drinking water is an essential need for sustaining human-well being. The majority of people suffering from lack of clean water reside in less developed communities, where the financial and technical barriers hamper the implementation of modern water-sanitation systems. The goal of this study is to evaluate hydrochar derived from selected organic wastes as a low cost adsorbent for pathogen removal in water treatment. Pathogen removal efficiency was measured by carrying out breakthrough analysis using a simple 10 cm sand filtration set-up supplemented with hydrochar (1.5 % *w/w*).

Two home-made two-step reverse transcription-quantitative polymerase chain reaction assays were developed using either RevertAid or MMLV reverse transcriptase. Both assays showed competitive efficiency to a selected commercial one-step master kit on analyzing environmental and laboratory samples. The costs of home-made assays were 11 times less than the commercial kit. Successive virus removal experiments were carried out using the home-made assay based on the RevertAid reverse transcriptase.

We evaluated the *Escherichia coli* (*E. coli*) removal efficiency of hydrochars derived from agricultural residue of maize and stabilized sewage sludge from wastewater treatment plant. Though raw hydrochars showed relatively limited *E. coli* removal efficiency of ~70 % for maize-hydrochar and ~20 % for sewage sludge-hydrochar, a simple cold-alkali activation of hydrochars using 1M KOH solution enhanced *E. coli* removal efficiency of both hydrochars (~90 %). Apparently, KOH activation of raw hydrochar removed alkali-soluble and tar-like substances from hydrochar surface, exposing more hydrophobic core and developing more porous surface structure. Also, removal of pathogenic rotavirus and adenovirus were investigated using hydrochar produced from sewage sludge and swine waste. Raw hydrochars (without activation) successfully removed both viruses at efficiencies higher than 99 %. Successive material characterization results suggested that the supplements of the hydrochar adsorbents provided a larger contact surface area with strong hydrophobicity in the sand media, facilitating adsorptive removal of pathogenic microorganisms.

This PhD study suggests the potential of hydrothermal carbonization as a technical solution for water-sanitation issues. Through hydrothermal carbonization, hazardous solid wastes containing pathogenic microorganisms will be totally sanitized, and the resulting hydrochars can be used as a pathogen barrier in water and / or wastewater treatment systems.

Samenvatting

Voldoende schoon drinkwater is een essentiële basisbehoefte voor de mens. De meerderheid van de wereldbevolking, die geen of onvoldoende toegang heeft tot schoon (drink)water woont in minder bedeelde gemeenschappen, waar financiele en technische barrières de implementatie van moderne water- en sanitatie systemen verhinderen. Het doel van deze studie is om te evalueren of zgn. 'hydrochar' of kortweg 'char', dat gemaakt wordt van organisch afval, een goed en goedkoop middel is om pathogenen te verwijderen uit afval water. De pathogene verwijderings efficientie is bepaald door middel van de analyse van doorbraakcurven in 10 cm zand kolommen, vermengd met char (1.5% g/g).

Twee zelf gemaakte twee-staps reverse transcription polymerase chain reaction protocollen zijn ontwikkeld gebaseerd op het gebruik van RevertAid danwel MMLV reverse transcriptase. In het analyseren van monsters uit het veld en het laboratorium werkten beide protocollen ongeveer even goed als een commerciele master kit. Echter, de analyse kosten waren bij de zelf gemaakte protocollen 11 keer lager dan bij gebruik van de commerciele master kit. Alle monsters van de virus verwijderings experimenten zijn vervolgens met de zelf gemaakte protocollen geanalyseerd.

In eerste instantie is de verwijderings efficientie van *Escherichia coli* bij gebruik van chars verkregen uit mais resten en gestabiliseerd zuiveringsslib geevalueerd. De onbewerkte chars lieten relatief weinig verwijdering zien van ongeveer 70 en 20%, respectievelijk. Echter, een simpele activatie van de chars door middel van wassen met 1 M KOH de *E. coli* verbeterde de verwijderings efficientie tot 90%. Waarschijnlijk verwijderde het wassen de alkali-oplosbare en teerachtige bestanddelen van het oppervlak van de chars, waardoor niet alleen meer hydrofoob oppervlak beschikbaar kwam, maar ook meer porien in de chars.
Ook is gekeken naar de verwijdering van pathogeen rotavirus en adenovirus uit water door chars gemaakt van zowel zuiveringsslib als van varkensmest. De verwijderings efficienties van de onbehandelde niet geactiveerde chars was meer dan 99%. Dit bleek te worden veroorzaakt door een groot en hydrofoob hechtings oppervlak van de chars, waardoor verwijdering door hechting plaats kon vinden.

Deze PhD studie laat zien, dat hydrothermale carbonisatie een potentieel goede technische oplossing kan bieden voor het water en sanitatie vraagstuk. Door middel van hydrothermale carbonisatie kan pathogeen-rijk organisch afval geheel worden gesaneerd en de geproduceerde chars kunnen gebruikt worden als barrière tegen pathogenen in (afval)water zuiverings installaties

Contents

Chapter 1: General introduction

1.1 Background

Clean water is an essential life-sustaining element that enables hygienic practices and secured drinking-water consumption. Approximately 10 percent of the total burden of diseases could be prevented through proper drinking water, sanitation, hygiene and water resource management (WHO, 2008). In 2002, the World Health Organization (WHO) reported that lack of safe water and insufficient sanitation practices attribute to approximately 1.7 million deaths and 54.2 million disability-adjusted life years (DALYs). Developing countries were classified as the most vulnerable regions occupying 99.8 % of deaths. Ninety percent of those deaths were little children less than 5 years old (WHO, 2002). In spite of the recent decrease of child mortality, water born diarrhoea still remains one of the main causes of child death. Complication of other diseases such as malaria and pneumonia stimulates diarrheal death of children (Kosek et al., 2003). Every year 500 million people are at the risk of Trachoma; among them 146 million are threatened by blindness and 6 million indeed lose their sight. Improvements of water accessibility and hygiene can reduce trachoma morbidity by 27 percent (WHO, 2004).

The international society has begun exerting efforts to remediate this tragic water-sanitation crisis. In 2000, the United Nations (UN) declared the Millennium Development Goals (MDGs) stating target 11 "Halve by 2015 the proportion of people without sustainable access to safe drinking water and basic sanitation" and target 12 "By 2020, to have achieved a significant improvement in the lives of at least 100 million slum dwellers" (WHO, 2006). Recently, the WHO reported progress of these efforts. Currently, 2.5 billion people are excluded from improved sanitation and 748 million people are living without improved drinking-water sources (WHO and UNICEF, 2014). Though the world population connected to improved drinking water and sanitation has increased, the number of people facing risks in a particular region also increased because of the rapid population growth (Aertgeerts, 2009).

The majority of people at risk of water borne diseases are distracted from sufficient public services. Rapid development of peri-urban areas contributes this situation. From the mid 20th century, the population in agricultural areas in developing countries declined by 20 to 30 % by migration to urban areas (UN-HABITAT, 2007). Nearly one billion people inhabited slum regions in 2003 and this number is projected to be doubled by 2030 (UN-Habitat, 2003). Although insufficient connection to sanitation facilities and poor safe water supply are strongly related to the high infant mortality and poor nutritional status of children (Checkley et al., 2004; Vaid et al., 2007), the high cost of installing piped water connections is almost unaffordable for the inhabitants with an income less than $2 per day (Kayaga and Franceys, 2007). As a temporary solution to minimize the health risks until providing proper sanitation facilities and safe water connection, the development of low cost technologies is necessary. Regarding the fact that the most vulnerable communities are also limited by technical and financial resources, the best solution enabling sustainable implementation would be low cost technology based on local resources with easy accessibility.

Hydrothermal carbonization (HTC) is a carbonizing process of biomass under wet conditions and low temperature between 180 and 200 °C (Titirici et al., 2007a). In hydrothermal processes, the solid material is surrounded by water during the reaction, which is kept in a liquid state by allowing the pressure to rise with the steam pressure in (high)-pressure reactors. Along with the increased research interests, HTC has been recommended as a carbon retention method mitigating green house effect (Titirici et al., 2007b; Titirici et al., 2008) and transformation method of biomass into valuable materials (Rillig et al., 2010; Ryu et al., 2010; Kumar et al., 2011; Mumme et al., 2011) having wide potential applications in water treatments, electronics, soil amendments and medicine (Titirici et al., 2007a). However, intensive investigations of HTC materials in water treatment for microbial contaminant removal have not been performed yet. This Ph. D thesis will evaluate the efficiency of HTC materials derived from selected organic wastes in water treatment mainly focusing on virus removal.

1.2   Research scope and objectives

This research aimed at providing evaluation of adsorbents produced via HTC of organic waste in water treatment for the removal of microorganism. The removal of test microorganisms were investigated by performing breakthrough analyses in simple sand filtration set-up supplemented with selected adsorbents derived from organic wastes. In order to fulfil this aim, specific objectives are formulated:

1. To develop home-made low cost reverse transcription quantitative polymerase chain reaction (RT-qPCR) / quantitative polymerase chain reaction (qPCR) assays for quantitative detection of rotavirus and adenovirus

2. To evaluate the removal efficiency of E. coli and / or pathogenic viruses in sand columns supplemented with hydrochar derived from selected organic waste (maize, sewage sludge or faecal waste).

3. To provide additional understanding on attachment - detachment behaviour of test microorganisms onto hydrochar surfaces.

1.3   Thesis outline

This thesis consists of eight chapters (see Figure 1.1). Chapter 1 presents a brief introduction of the study. Chapter 2 gives an in-depth literature review on the main components of the research. Chapter 3 provides development of low cost (RT)-qPCR protocols for rotavirus quantification in water samples. Chapter 4 demonstrates the Escherichia coli (E. coli) removal performance of hydrochar derived from agricultural waste of maize. Chapter 5 presents a study on the application of sewage sludge-derived hydrochar for *E. coli* removal in sand filtration set-ups with intermittent operation for 30 days. Chapter 6 shows the results of pathogenic virus removal using the hydrochar derived from sewage sludge. The effect of humic acid in viral attach-detachment was investigated. Chapter 7 provides an evaluation on hydrochar derived from faecal waste as an adsorbent for virus removal in water treatment. Simultaneous removal of pathogenic rotavirus and adenovirus was investigated. Chapter 8 consists of a general discussion, conclusions and recommendations for future research.

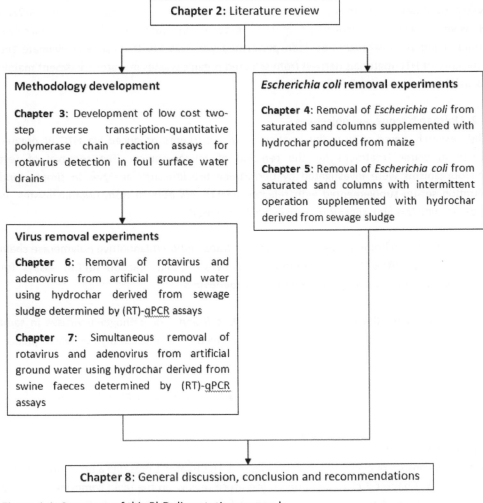

Figure 1.1: Structure of this PhD dissertation research

1.4 References

Aertgeerts, R. (2009) Progress and challenges in water and sanitation. Desalination 248(1-3), 249-255.

Kosek, M., Bern, C. and Guerrant, R.L. (2003) The global burden of diarrhoeal disease, as estimated from studies published between 1992 and 2000. Bulletin of the World Health Organization 81(3), 197-204.

Kumar, S., Kothari, U., Kong, L., Lee, Y.Y. and Gupta, R.B. (2011) Hydrothermal pretreatment of switchgrass and corn stover for production of ethanol and carbon microspheres. Biomass and Bioenergy 35(2), 956-968.

Mumme, J., Eckervogt, L., Pielert, J., Diakité, M., Rupp, F. and Kern, J. (2011) Hydrothermal carbonization of anaerobically digested maize silage. Bioresource Technology 102(19), 9255-9260.

Rillig, M.C., Wagner, M., Salem, M., Antunes, P.M., George, C., Ramke, H.-G., Titirici, M.-M. and Antonietti, M. (2010) Material derived from hydrothermal carbonization: Effects on plant growth and arbuscular mycorrhiza. Applied Soil Ecology 45(3), 238-242.

Ryu, J., Suh, Y.-W., Suh, D.J. and Ahn, D.J. (2010) Hydrothermal preparation of carbon microspheres from mono-saccharides and phenolic compounds. Carbon 48(7), 1990-1998.

Titirici, M.-M., Antonietti, M. and Baccile, N. (2008) Hydrothermal carbon from biomass: a comparison of the local structure from poly- to monosaccharides and pentoses/hexoses. Green Chemistry 10(11), 1204-1212.

Titirici, M.-M., Thomas, A. and Antonietti, M. (2007b) Back in the black: hydrothermal carbonization of plant material as an efficient chemical process to treat the CO2 problem? New Journal of Chemistry 31(6), 787-789.

Titirici, M.M., Thomas, A., Yu, S.-H., Mueller, J.-O. and Antonietti, M. (2007a) A direct synthesis of mesoporous carbons with bicontinuous pore morphology from crude plant material by hydrothermal carbonization. Chemistry of Materials 19(17), 4205-4212.

WHO (2002) World Health Report 2002 'Reducing Risks, Promoting Healthy Life'.

WHO (2004) Water, Sanitation and Hygiene Links to Health, 'FACTS AND FIGURES'.

WHO (2006) PREVENTING DISEASE THROUGH HEALTHY ENVIRONMENTS, 'Towards an estimate of the environmental burden of disease'.

WHO (2008) Guidelines for drinking-water quality, third edition, incorporating first and second addenda.

WHO and UNICEF (2014) Progress on drinking water and sanitation: 2014 update, WHO, Geneva, Switzerland / UNICEF, New York, USA. ISBN: 978-92-4-150724-0.

Chapter 2: Literature review

2.1 Hydrothermal carbonization

Hydrothermal carbonization is a thermo-chemical conversion process of biomass that aims to yield useful materials. Valuable carbonaceous products can be produced by heating a suspension of feed stock and water at temperatures of 180–200°C at saturated pressure for several hours (Funke and Ziegler, 2010). In contrast to the conventional dry pyrolysis, HTC can convert wet-biomass into carbonaceous materials without an energy-intensive drying before or during the process. In addition to the economic advantage obtained from less fuel consumption, applicability of wet biomass initiated the research on various non-traditional potential feed stocks: wet animal / human excrement, sewage sludges, municipal solid waste, aquaculture and algal residues (Libra et al., 2011).

The complex reaction network of HTC is not known in detail yet. Among many chemical reactions that might occur in HTC dehydration and decarboxylation have been focussed governing carbonization process inducing exothermal reaction that decreases the oxygen and hydrogen content (Peterson et al., 2008). The quality of HTC product is determined by nature of feed stock and several process parameters such as hydrous condition, temperature, residence time, pressure, solid load and pH. In comparison to conventional dry pyrolysis HTC requires longer residence times ranging from a few hours to several days. Significant increase of HTC coal yield has been observed with longer residence time. It can be explained by polymerization of solved fragments in liquid suspension that leads precipitation of insoluble solids in elongated reaction (Schuhmacher et al., 1960; Sevilla and Fuertes, 2009). Water has a decisive role in HTC. In natural system the importance of hot water as a reactant, solvent and catalyst for organic compounds is well recognized (Siskin and Katritzky, 2001). The solvent property of water is significantly improved at high temperature, and it becomes relevant even for non-polar compounds. It facilitates the transportation of biomass fragments out of the structural matrix which isolates these fragments from the reaction (Ross et al., 1991; Michels and Landais, 1994). In addition water is a good medium for heat transfer and storage. In wet condition pyrolysis that might result from local temperature elevation in HTC is inhibited by surrounding water (Funke and Ziegler, 2010). Alike conventional carbonization process HTC produces solids, liquids and gases. HTC yields more solid products, more water soluble organic compounds and less gases (mainly CO_2) than dry pyrolysis (Yao et al., 2007). Generally the chemical structure of HTC product is more similar to natural coal than dry pyrolysis products (Schuhmacher et al., 1960; Sugimoto and Miki, 1997). Among various applications of HTC products, main research flows in water treatment has been given to heavy metal removal. It was reported that HTC can remove heavy metal from water without additional activation process which is an essential requirement for dry pyrolysis products. Materials with highly functional surface area were produced by HTC via incorporation of very small amounts of carboxylic groups containing organic monomers in the carbon structure. HTC adsorbent achieved higher capacities than standard synthetic ion exchange resins and other types of sorption materials in cadmium and lead removal (Demir-Cakan et al., 2009).

To our knowledge, intensive research about the application of HTC materials for pathogen removal in water treatment has not been published yet. This research focuses on a role of virus adsorbent in water treatment. The size of significant human pathogenic viruses resulting in water borne diseases ranges from 20 to 100 nm in diameter (WHO, 2008). Regarding this size distribution, HTC materials can provide meso-macroporous adsorption sites (cf. mesopore 2 ~ 50 nm, macropore > 50 nm) (Figure 2.1) for viruses removal in water treatment that can be implemented by relatively easy technology with low cost (Titirici et al., 2007).

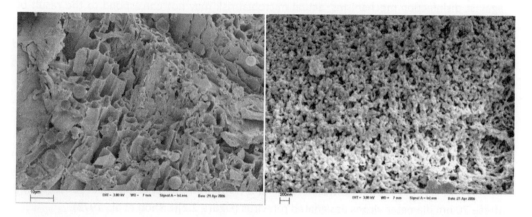

Figure 2.1: (a) Low-magnification SEM overview of a HTC treated oak leaf; scale bar 10 *i*m; (b) high-magnification picture (scale bar, 200 nm) of the same, sample indicating major changes of the nanostructure and the creation of a mesopore transport system due to chemical dehydration of the scaffold (Titirici et al., 2007).

2.2   Test microorganisms

2.2.1   *Escherichia coli*

Escherichia coli (*E. coli*) is a gram-negative, facultative anaerobic bacterium that belongs to the family of *Enterobacteriaceae* (Foppen and Schijven, 2006). *E. coli* is recognized as the most important parameter of fecal contaminations by microbiology and public health experts. (Bartram et al., 2004). In all mammal faeces, it is found with concentration of ca 10^9 / ml and does not propagate well in the environment. Depending on environmental conditions, *E. coli* can survive for 4 to 12 weeks (Edberg et al., 2000). There are various factors affecting the survival of *E. coli* in environments such as protozoa, antagonists, temperature, light, soil, pH, toxic substances and oxygen (Foppen and Schijven, 2006). The survival periods of *Escherichia coli* in various surroundings were reported; in the ground water at 10°C, recharged well and river water at 9-16 °C, *Escherichia coli* survived for 100 days, 63 days and 55 days respectively (Goldshmi.Mg and Pantsili.Vd, 1972; Grabow et al.,

1975; Filip et al., 1987). Due to its strong relevance with the faecal contamination and relatively easy quantification methods, *E. coli* has been employed in wide range of investigations including water treatment (Muhammad et al., 2008; Bielefeldt et al., 2009; van Halem et al., 2009; Fisher et al., 2012) and transport of faecal contamination in aquifer (Wong et al., 2008; Lutterodt et al., 2011; Sinton et al., 2012).

Although the concept of indicator organism is gradually accepted, there exists certain limitation. Waterborne diseases are not only caused by bacterial pathogen but may result from protozoan parasites and viruses. Because these groups of pathogen have different survival characteristics and behavior related to environmental stresses and resistance against disinfection mechanisms, actual microbial risk may not correspond to the result of faecal indicator organism assessment. This must be considered in microbial risk assessment and water treatment technology development.

2.2.2 Rotavirus

Rotavirus is one of the most significant human pathogenic viruses resulting in acute gastroenteritis responsible for more than 527,000 annual deaths. Worldwide, 95% of children are infected by rotavirus from 3 to 5 years of age in both developed and developing countries (Parashar et al., 1998; Parashar et al., 2009). The first observation of rotavirus was reported in 1973. Bishop and his colleagues visualized unique viral particles by electron microscopy from duodenal epithelium of children with non-bacterial diarrhea. Subsequently these 70 nm diameter virons designated rotavirus (Figure 2.2)(Bishop et al., 1973).

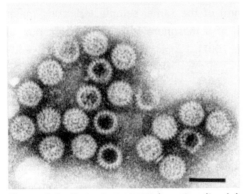

Figure 2.2: Rotavirus particles visualized by immune electron microscopy in stool filtrate from child with acute gastroenteritis. 70-nm particles possess distinctive double-shelled outer capsid. Bar = 100 nm. (Parashar et al., 1998)

Rotavirus is a non-enveloped icosahedral virus classified within the *Reoviridae* family. The core of virion contains virus genome comprising 11 segments of double stranded ribonucleic acid (dsRNA). Each RNA encodes either viral proteins producing the viral capsid or non-structural proteins (NSP). The viral capsid proteins determine the group, subgroup, and serotype of rotavirus. Among seven major groups (A-G), group A causes most human infections, while groups B and C only associated occasional human illness (Parashar et al., 1998). Rotavirus transmits mainly through fecal-oral routes; unwashed hands, focally contaminated water and food are recognized as main carriers. Despite of the observation of rotaviruses in urine and upper-respiratory samples, these body fluids are not regarded to be associated to transmission (Vollet et al., 1981; Zheng et al., 1991). After the ingestion, virons are delivered to small intestine where the viral replication occurs. The strong resistance against acidic condition in range of human gastric pH enables transmission of rotavirus through gastrointestinal tract. More than 10^{10} - 10^{11} rotavirus particles were contained in 1 gram of fecal excrement of infected children (Desselberger, 1999). The presence of rotavirus in surface water, waste water and biosolids has been reported. The icosahedral shell structure of rotavirus induces very strong resistance against environmental stress (Gerba et al., 1996a; Chapron et al., 2000; Estes et al., 2008). In surface water rotavirus could survive 8 to 32 days; also it is one of most tolerant enteric virus to ultraviolet radiation and chloramines (Gerba et al., 1996b).

2.2.3 Adenovirus

Human adenoviruses cause wide range of infections in the gastrointestinal tract (gastroenteritis), the respiratory tract (acute respiratory diseases, pneumonia, pharyngoconjunctival fever) and the urinary tract. More than 80% of infections occur in children under 4 years old due to insufficient immunity development. It is the third most important source of gastroenteritis in children after rotavirus and norovirus notably in developing communities (Gomara et al., 2008).

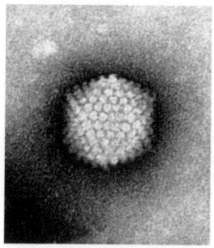

Figure 2.3: Adenovirus after negative stain electron microscopy (Gelderblom, 1996)

Human adenovirus is a non-enveloped double-stranded deoxyribonucleic acid (dsDNA) virus. Linear dsDNA with around 35kilo bases encoding more than 30 structural and non-structural proteins embedded in icosahedral capsid with size of 90 to 100 nm in diameter (Figure 2.3)(Friefeld et al., 1984; Stewart et al., 1993). Adenovirus belonged to the genus *Mastadenovirus* in the *Adenoviridae* family (Lu and Erdman, 2006; Henquell et al., 2009). Seven species (A to G) and Fifty-two serotypes of adenovirus are classified (Jones et al., 2007; Kajon et al., 2010; Lee et al., 2010). Though species A, B, C, D, and E are observed worldwide and have been associated in outbreaks in humans, more than half of serotypes are infrequently detected (Louie et al., 2008) and only one-third of serotypes are responsible for human disease (Lynch et al., 2011). Serotype Ad40 and Ad41 in species F are reported being responsible for majority of adenovirus associated gastroenteritis in children. (Lynch et al., 2011). The worldwide prevalence of adenovirus is

11

observed in river, coastal waters, swimming pools and treated drinking waters (Jiang, 2006). Most serotypes are persistent in the environment both on dry inanimate surface (Kramer et al., 2006) and in aquatic conditions including tap water, sewage effluent and sea water (Enriquez et al., 1995). Also adenovirus has stronger resistance to UV radiation than other enteric viruses (Thurston-Enriquez et al., 2003).

2.3 Polymerase chain reaction technologies

In this research, the reverse transcription quantitative polymerase chain reaction (RT-qPCR) technology is selected to investigate the pathogenic viruses. RT-qPCR is an outstanding molecular biological tool in various research fields. In the microbial water quality analysis it is regarded as the golden standard enabling sensitive quantitative and qualitative analysis. The concept of quantitative PCR (qPCR) is an amplification of selected deoxyribonucleic acid (DNA) fragment. The amplification of target template is estimated by measurement of a fluorescence signal. For analysis of target ribonucleic acid (RNA) a reverse transcription (RT) step is necessary prior to conventional qPCR. RT is conducted by the enzyme reverse transcriptase (RTase), which converts RNA to complementary deoxyribonucleic acid (cDNA). This cDNA can be amplified in a subsequent qPCR reaction for detection. RTase can be obtained from moloney murine leukemia virus (MuLV)(Roth et al., 1985) and avian myeloblastosis virus (AMV) (Houts et al., 1979). The majority of commercial RTase are produced from these two sources. The sensitivity of some RT-qPCR reactions can be affected by the type of RTase (BarraganGonzalez et al., 1997). Several researches evaluated the performance of various commercial RTases focusing on resistance to common RT-qPCR inhibitors, detection limits in low/high amounts of background RNA and sensitivities (Levesque-Sergerie et al., 2007; Arezi et al., 2010; Okello et al., 2010). Although these reports can be a good background information, there still exists a need for research into RTase selection, depending on specific research interests.

RT-qPCR / qPCR technologies are outstanding molecular biological tools in various research fields. In the microbial water quality analysis it is regarded as the golden standard enabling sensitive quantitative and qualitative analysis. The concept of qPCR can simply be explained as amplification of selected deoxyribonucleic acid (DNA) fragment. It is carried out in reaction mix that consists of thermostable DNA polymerase, dNTPs (free nucleotides that will be composed in DNA synthesis), gene-specific primers, reporter (SYBR green or Taqman probe) and nucleic acid extract from sample to be analyzed. Within several stages of thermal cycling following reactions occur in reaction mix.

- Denaturation - breaking down of the dsDNA of the sample by melting at high temperature normally 94-96°C
- Annealing - attachment of gene-specific oligonucleotide primers to the denatured DNA

- Extension - elongation of the attached primers by the DNA polymerase and dNTPs resulting in a copy of DNA template
- Plate reading – measurement of fluorescenic signals from sample

In ideal status the number of target nucleotide sequence (template) doubles every cycle resulting in increase of fluorescence emission. Due to the fact that the strength of signals from reporter is directly proportional to the initial amount of target template in the sample, quantitative measurement can be made by regression analysis. This is normally done by built in software of thermal cycler. For analysis of target RNA a reverse transcription (RT) step is necessary prior to conventional qPCR. RT is conducted by the enzyme reverse transcriptase (RTase), which converts RNA to complementary deoxyribonucleic acid (cDNA). This cDNA can be amplified in a subsequent qPCR. RTase can be obtained from moloney murine leukemia virus (MuLV)(Roth et al., 1985) and avian myeloblastosis virus (AMV) (Houts et al., 1979). The majority of commercial RTase are produced from these two sources. The sensitivity of some RT-qPCR reactions can be affected by the type of RTase (BarraganGonzalez et al., 1997). Several researches evaluated the performance of various commercial RTases focusing on resistance to common RT-qPCR inhibitors, detection limits in low / high amounts of background RNA and sensitivities (Levesque-Sergerie et al., 2007; Arezi et al., 2010; Okello et al., 2010). Although these reports can be a good background information, there still exists a need for the research into RTase selection depending on specific research interests.

Several variations of the PCR technology have been developed in order to overcome technical and economical limitations. Hot start PCR is used to reduce non-targeted amplification during the initial stages of PCR. It could be performed by denaturing reaction components before the addition of polymerase. In addition, specialized enzyme systems have been developed that inhibit the polymerase's activity at ambient temperature, either by the binding of an antibody or by the presence of covalently bound inhibitors that dissociate only after a high-temperature activation step (Chou et al., 1992).

Nested PCR uses two pairs of primers in order to increase the specificity of DNA amplification by minimizing non-specific amplification of DNA. Two successive PCRs are performed with two sets of primers. In the first reaction, one pair of primers is used to generate DNA products which may still consist of non-specifically amplified DNA fragments besides the intended target. The products are then used in a second PCR with a set of primers whose binding sites are completely or partially different. Nested PCR usually has a higher efficiency on amplification of long DNA fragments than conventional PCR. In order to have successful nested PCR, more detailed knowledge of the target sequences and precise primer design are crucial necessities. It has been used in various research fields including Human Picornavirus and enteric virus detection from polluted water (Kammerer et al., 1994; Puig et al., 1994). The application of PCR technologies could be limited by cost and the availability of an adequate test sample volume. To overcome these infirmities multiplex PCR has been developed (Bej et al., 1990). In multiplex PCR more than one target nucleic acid sequence can be amplified by application of more than one pair of primers in the same reaction. Multiplex PCR has the potential of minimizing time consumption and laboratory works. It can be used not only for detection but also for the genetic characterization of target microorganisms (Paton and Paton, 1998); Yet, the optimization of multiplex PCR would be difficult. Combination with the non-mechanical hot-start PCR would be recommended (Elnifro et al., 2000). Integrated cell culture PCR (ICC-PCR) is a combination of cell culture and PCR technology. This is developed to remediate individual limitations of cell culture and PCR. PCR analysis can provide detection of the template at a very sensitive level. The concept of cell culture can dilute away PCR inhibitors in samples. Also, the cultivation of target microorganisms provides enhanced sensitivity of PCR assays as well as information about the viability / infectivity of target microorganisms. Several researches evaluating ICC-PCR reported better efficiency and sensitivity than either conventional PCR or cell culture methods alone (Blackmer et al., 2000; Chapron et al., 2000; Jiang et al., 2004; Lee et al., 2005).

The main limitation of the qPCR technology on the evaluation of microbial water quality can be the inability of distinguishing live and dead microorganism. Due to the fact that DNA is very persistent in the environment, residual DNA from inactivated cells can also give positive signals in qPCR, leading to an over-estimation of remaining active microorganisms (Josephson et al., 1993; Masters et al., 1994). Because of the infectivity of viruses is determined by property of viral membrane that has an essential role of attachment on host cell and integrity of nucleic acid containing genetic instruction, same concept of the false positive result is applied to qPCR technology on viruses.

2.4   Mechanisms for colloid retention in porous media

The initial attachment of microbial colloids onto solid surfaces in liquid phase in controlled laboratory conditions could be generally considered to be similar to colloidal deposition to which the classic Derjaguin-Landau-Verwey-Overbeek (DLVO) theory has been applied (Redman et al., 2004). DLVO theory describes the inter-surface potential energy by summing the electrostatic (double-layer) repulsion or attraction, London–van der Waals attraction and short-range forces such as hydration and steric repulsion (Figure 2.4)(Ryan and Elimelech, 1996). The net potential energy curve comprises an attractive energy well (primary minimum, ϕ_{min1}) at a very short separation distance (δ), a repulsive energy barrier (ϕ_{max}), and a shallow attractive energy well at a larger separation distance (secondary minimum, $\phi min2$) (Schijven and Hassanizadeh, 2000).

Under most environmental conditions, both biological colloids and media surfaces (silicate mineral grains) have net-negative surface charge suggesting repulsive electrostatic forces

(Davis, 1982; Tipping and Cooke, 1982). In these conditions that DLVO predicts unfavourable attachment of colloids, several phenomena have been suggested as potential retention mechanisms such as deposition in the secondary energy minimum, hydrophobic interaction, straining and heterogenic charge distribution.

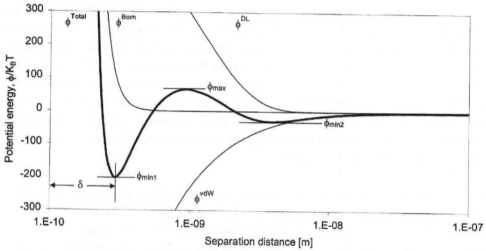

Figure 2.4: DLVO energy as a function of separation distance between a colloid and a collector. The total potential energy (ϕ^{Total}) is the sum of the double layer potential energy (ϕ^{DL}), the van der Waals potential energy (ϕ^{vdW}), and the Born potential energy (ϕ^{Born}). The total potential energy curve is characterized by an attractive well at a very small separation distance (δ), the primary minimum (ϕ_{min1}), a repulsive energy barrier (ϕ_{max}), and a shallow attractive well at a larger separation distance (ϕ_{min2}). The potential energy is normalized by kBT (Ryan and Elimelech, 1996).

The deposition of colloids in the secondary energy minimum was suggested as an important attachment mechanism under unfavourable conditions (Hahn and O'Melia, 2004). In contrast to the deposition in the primary energy minimum that has been considered as irreversible attachment, the colloids associated in the secondary energy minimum was detached from the collector surface by decreasing the ionic strength, leading to an increase in the double layer repulsion (Walker et al., 2004; Foppen et al., 2007).

It was suggested that the hydrophobic interaction plays an important role in virus attachment on solid media when both surfaces are negatively charged (Gerba, 1984; Bales et al., 1991). Bales (1993) et al., suggested that the importance of hydrophobic effects in an attachment of bacteriophage MS-2 in soil sediment may be orders of magnitude more important than electrostatic forces. Hydrophobic interactions resulted from the thermodynamically unfavourable interaction of hydrophobic substances with water molecules and are not due to the interaction between hydrophobic surfaces themselves

(Schijven and Hassanizadeh, 2002). The association of a hydrophobic particle with another hydrophobe or preferential attachment site is favoured thermodynamically owing to less need for reordering of water molecules than when a hydrophobic particle remains suspended in liquid (Wait and Sobsey, 1983).

Straining of colloids during the transportation through porous media involves the entrapment of colloids in pores or grain junctions that are small enough to exclude the colloids from fluid streamlines (Bradford et al., 2004; Johnson et al., 2007). The straining process is largely determined by the threshold ratio that represents the ratio of the colloid diameter to median grain diameter. Both simulation and laboratory experiments indicated that straining played an important role in the retention of colloids when the threshold ration is greater than 0.005 (Bradford et al., 2004; Johnson et al., 2007). Besides the threshold ratio, irregularity of grain shape significantly contributed the straining potential of the porous media (Tufenkji et al., 2004).

Also, the colloidal deposition in unfavourable condition could be attributed to heterogenic surface charge distribution of the media surface (Truesdail et al., 1998; Ryan et al., 1999). Despite net-negative surface charge of media surface, positively or less negatively charged patches would provide local zones with favourable conditions for colloidal deposition (Elimelech et al., 2000).

2.5 References

Arezi, B., McCarthy, M. and Hogrefe, H. (2010) Mutant of Moloney murine leukemia virus reverse transcriptase exhibits higher resistance to common RT-qPCR inhibitors. Analytical Biochemistry 400(2), 301-303.

Bales, R.C., Hinkle, S.R., Kroeger, T.W., Stocking, K. and Gerba, C.P. (1991) Bacteriophage adsorption during transport through porous media: chemical perturbations and reversibility. Environmental Science & Technology 25(12), 2088-2095.

Bales, R.C., Li, S., Maguire, K.M., Yahya, M.T. and Gerba, C.P. (1993) MS-2 and poliovirus transport in porous media: Hydrophobic effects and chemical perturbations. Water Resources Research 29(4), 957-963.

BarraganGonzalez, E., LopezGuerrero, J.A., BoluferGilabert, P., SanzAlonso, M., DelaRubiaComos, J. and SempereTalens, A. (1997) The type of reverse transcriptase affects the sensitivity of some reverse transcription PCR methods. Clinica Chimica Acta 260(1), 73-83.

Bartram, J., Cotruvo, J., Exner, M., Fricker, C. and Glasmacher, A. (2004) Heterotrophic plate count measurement in drinking water safety management - Report of an Expert Meeting Geneva, 24-25 April 2002. International Journal of Food Microbiology 92(3), 241-247.

Bej, A.K., Mahbubani, M.H., Miller, R., DiCesare, J.L., Haff, L. and Atlas, R.M. (1990) Multiplex PCR amplification and immobilized capture probes for detection of bacterial pathogens and indicators in water. Molecular and Cellular Probes 4(5), 353-365.

Bielefeldt, A.R., Kowalski, K. and Summers, R.S. (2009) Bacterial treatment effectiveness of point-of-use ceramic water filters. Water Research 43(14), 3559-3565.

Bishop, R., Davidson, G.P., Holmes, I.H. and Ruck, B.J. (1973) VIRUS PARTICLES IN EPITHELIAL CELLS OF DUODENAL MUCOSA FROM CHILDREN WITH ACUTE NON-BACTERIAL GASTROENTERITIS. The Lancet 302(7841), 1281-1283.

Blackmer, F., Reynolds, K.A., Gerba, C.P. and Pepper, I.L. (2000) Use of integrated cell culture-PCR to evaluate the effectiveness of poliovirus inactivation by chlorine. Applied and Environmental Microbiology 66(5), 2267-2268.

Bradford, S.A., Bettahar, M., Simunek, J. and van Genuchten, M.T. (2004) Straining and attachment of colloids in physically heterogeneous porous media. Vadose Zone Journal 3(2), 384-394.

Camesano, T.A. and Logan, B.E. (1998) Influence of fluid velocity and cell concentration on the transport of motile and nonmotile bacteria in porous media. Environmental Science & Technology 32(11), 1699-1708.

Chapron, C., D., Ballester, N., A., Fontaine, J., H., Frades, C., N. and Margolin, A., B. (2000) Detection of astroviruses, enteroviruses, and adenovirus types 40 and 41 in surface waters collected and evaluated by the information collection rule and an integrated cell culture-nested PCR procedure. Applied and Environmental Microbiology 66(6), 2520-2525.

Chou, Q., Russell, M., Birch, D.E., Raymond, J. and Bloch, W. (1992) Prevention pre-PCR MIS-priming and primer dimerization improves low-copy number amplifications. Nucleic Acids Research 20(7), 1717-1723.

Davis, J.A. (1982) Adsorption of natural dissolved organic matter at the oxide/water interface. Geochimica et Cosmochimica Acta 46(11), 2381-2393.

Demir-Cakan, R., Baccile, N., Antonietti, M. and Titirici, M.M. (2009) Carboxylate-Rich Carbonaceous Materials via One-Step Hydrothermal Carbonization of Glucose in the Presence of Acrylic Acid. Chemistry of Materials 21(3), 484-490.

Desselberger, U. (1999) Rotavirus infections - Guidelines for treatment and prevention. Drugs 58(3), 447-452.

Edberg, S.C., Rice, E.W., Karlin, R.J. and Allen, M.J. (2000) Escherichia coli: the best biological drinking water indicator for public health protection. Journal of Applied Microbiology 88, 106s-116s.

Elimelech, M., Nagai, M., Ko, C.H. and Ryan, J.N. (2000) Relative insignificance of mineral grain zeta potential to colloid transport in geochemically heterogeneous porous media. Environmental Science & Technology 34(11), 2143-2148.

Elnifro, E.M., Ashshi, A.M., Cooper, R.J. and Klapper, P.E. (2000) Multiplex PCR: Optimization and application in diagnostic virology. Clinical Microbiology Reviews 13(4), 559-+.

Enriquez, C.E., Hurst, C.J. and Gerba, C.P. (1995) Survival of the enteric adenoviruses 40 and 41 in tap, sea, and waste water. Water Research 29(11), 2548-2553.

Estes, M.K., Kang, G., Zeng, C.Q.Y., Crawford, S.E. and Ciarlet, M. (2008) Gastroenteritis Viruses, pp. 82-100, John Wiley & Sons, Ltd.

Filip, Z., Kaddumulindwa, D. and Milde, G. (1987) Survival and Adhesion of Some Pathogenic and Facultative Pathogenic Microorganisms in Groundwater. Water Science and Technology 19(7), 1189-1189.

Fisher, M.B., Iriarte, M. and Nelson, K.L. (2012) Solar water disinfection (SODIS) of Escherichia coli, Enterococcus spp., and MS2 coliphage: Effects of additives and alternative container materials. Water Research 46(6), 1745-1754.

Foppen, J.W., van Herwerden, M. and Schijven, J. (2007) Transport of Escherichia coli in saturated porous media: Dual mode deposition and intra-population heterogeneity. Water Research 41(8), 1743-1753.

Foppen, J.W.A. and Schijven, J.F. (2006) Evaluation of data from the literature on the transport and survival of Escherichia coli and thermotolerant coliforms in aquifers under saturated conditions. Water Research 40(3), 401-426.

Friefeld, B.R., Lichy, J.H., Field, J., Gronostajski, R.M., Guggenheimer, R.A., Krevolin, M.D., Nagata, K., Hurwitz, J. and Horwitz, M.S. (1984) The in vitro replication of adenovirus DNA. Current topics in microbiology and immunology 110, 221-255.

Funke, A. and Ziegler, F. (2010) Hydrothermal carbonization of biomass: A summary and discussion of chemical mechanisms for process engineering. Biofuels Bioproducts & Biorefining-Biofpr 4(2), 160-177.

Gelderblom, H.R. (1996) Structure and Classification of Viruses. Galveston, U.o.T.M.B.a. (ed), University of Texas Medical Branch at Galveston, Baron.

Gerba, C.P. (1984) Advances in Applied Microbiology. Allen, I.L. (ed), pp. 133-168, Academic Press.

Gerba, C.P., Rose, J.B. and Haas, C.N. (1996a) Sensitive populations: Who is at the greatest risk? International Journal of Food Microbiology 30(1-2), 113-123.

Gerba, C.P., Rose, J.B., Haas, C.N. and Crabtree, K.D. (1996b) Waterborne rotavirus: A risk assessment. Water Research 30(12), 2929-2940.

Goldshmi.Mg and Pantsili.Vd (1972) Device for Carving out of Spacing Nets as Series of Circles. Zavodskaya Laboratoriya (1), 117-&.

Gomara, M.I., Simpson, R., Perault, A.M., Redpath, C., Lorgelly, P., Joshi, D., Mugford, M., Hughes, C.A., Dalrymple, J., Desselberger, U. and Gray, J. (2008) Structured surveillance of infantile gastroenteritis in East Anglia, UK: incidence of infection with common viral gastroenteric pathogens. Epidemiology and Infection 136(1), 23-33.

Grabow, W.O.K., Prozesky, O.W. and Burger, J.S. (1975) Behavior in a River and Dam of Coliform Bacteria with Transferable or Non-Transferable Drug-Resistance. Water Research 9(9), 777-782.

Grasso, D., Smets, B.F., Strevett, K.A., Machinist, B.D., VanOss, C.J., Giese, R.F. and Wu, W. (1996) Impact of physiological state on surface thermodynamics and adhesion of Pseudomonas aeruginosa. Environmental Science & Technology 30(12), 3604-3608.

Hahn, M.W. and O'Melia, C.R. (2004) Deposition and Reentrainment of Brownian Particles in Porous Media under Unfavorable Chemical Conditions: Some Concepts and Applications. Environmental Science & Technology 38(1), 210-220.

Harvey, R.W., Kinner, N.E., Bunn, A., Macdonald, D. and Metge, D. (1995) TRANSPORT BEHAVIOR OF GROUNDWATER PROTOZOA AND PROTOZOAN-SIZED MICROSPHERES IN SANDY AQUIFER SEDIMENTS. Applied and Environmental Microbiology 61(1), 209-217.

Henquell, C., Boeuf, B., Mirand, A., Bacher, C., Traore, O., Dechelotte, P., Labbe, A., Bailly, J.L. and Peigue-Lafeuille, H. (2009) Fatal adenovirus infection in a neonate and transmission to health-care workers. Journal of Clinical Virology 45(4), 345-348.

Houts, G.E., Miyagi, M., Ellis, C., Beard, D. and Beard, J.W. (1979) Reverse transcriptase from avian myeloblastosis virus. J Virol 29(2), 517-522.

Jiang, S.C. (2006) Human Adenoviruses in water: Occurrence and health implications: A critical review. Environmental Science & Technology 40(23), 7132-7140.

Jiang, Y.J., Liao, G.Y., Zhao, W., Sun, M.B., Qian, Y., Bian, C. and Jiang, S. (2004) Detection of infectious hepatitis A virus by integrated cell culture/strand-specific reverse transcriptase-polymerase chain reaction. Journal of Applied Microbiology 97(5), 1105-1112.

Johnson, W.P., Li, X. and Yal, G. (2007) Colloid Retention in Porous Media: Mechanistic Confirmation of Wedging and Retention in Zones of Flow Stagnation. Environmental Science & Technology 41(4), 1279-1287.

Jones, M.S., Harrach, B., Ganac, R.D., Gozum, M.M.A., dela Cruz, W.P., Riedel, B., Pan, C., Delwart, E.L. and Schnurr, D.P. (2007) New adenovirus species found in a patient presenting with gastroenteritis. Journal of Virology 81(11), 5978-5984.

Josephson, K.L., Gerba, C.P. and Pepper, I.L. (1993) Polymerase chain-reaction detection of nonviable bacterial pathogens. Applied and Environmental Microbiology 59(10), 3513-3515.

Jucker, B.A., Zehnder, A.J.B. and Harms, H. (1998) Quantification of polymer interactions in bacterial adhesion. Environmental Science & Technology 32(19), 2909-2915.

Kajon, A.E., Lu, X.Y., Erdman, D.D., Louie, J., Schnurr, D., St George, K., Koopmans, M.P., Allibhai, T. and Metzgar, D. (2010) Molecular Epidemiology and Brief History of Emerging Adenovirus 14-Associated Respiratory Disease in the United States. Journal of Infectious Diseases 202(1), 93-103.

Kammerer, U., Kunkel, B. and Korn, K. (1994) NESTED PCR FOR SPECIFIC DETECTION AND RAPID IDENTIFICATION OF HUMAN PICORNAVIRUSES. Journal of Clinical Microbiology 32(2), 285-291.

Kramer, A., Schwebke, I. and Kampf, G. (2006) How long do nosocomial pathogens persist on inanimate surfaces? A systematic review. Bmc Infectious Diseases 6.

Lee, J., Choi, E.H. and Lee, H.J. (2010) Comprehensive Serotyping and Epidemiology of Human Adenovirus Isolated From the Respiratory Tract of Korean Children Over 17 Consecutive Years (1991-2007). Journal of Medical Virology 82(4), 624-631.

Lee, S.H., Lee, C., Lee, K.W., Cho, H.B. and Kim, S.J. (2005) The simultaneous detection of both enteroviruses and adenoviruses in environmental water samples including tap water with an integrated cell culture-multiplex-nested PCR procedure. Journal of Applied Microbiology 98(5), 1020-1029.

Levesque-Sergerie, J.P., Duquette, M., Thibault, C., Delbecchi, L. and Bissonnette, N. (2007) Detection limits of several commercial reverse transcriptase enzymes: impact on the low- and high-abundance transcript levels assessed by quantitative RT-PCR. Bmc Molecular Biology 8.

Libra, J.A., Ro, K.S., Kammann, C., Funke, A., Berge, N.D., Neubauer, Y., Titirici, M., Fuhner, C., Bens, O., Kern, J. and Emmerich, K. (2011) Hydrothermal carbonization of biomass residuals: A comparative review of the chemistry, processes and applications of wet and dry pyrolysis. Biofuels 2:89-124.

Louie, J.K., Kajon, A.E., Holodniy, M., Guardia-LaBar, L., Lee, B., Petru, A.M., Hacker, J.K. and Schnurr, D.P. (2008) Severe pneumonia due to adenovirus serotype 14: A new respiratory threat? Clinical Infectious Diseases 46(3), 421-425.

Lu, X. and Erdman, D.D. (2006) Molecular typing of human adenoviruses by PCR and sequencing of a partial region of the hexon gene. Archives of Virology 151(8), 1587-1602.

Lutterodt, G., Foppen, J.W.A., Maksoud, A. and Uhlenbrook, S. (2011) Transport of Escherichia coli in 25 m quartz sand columns. Journal of Contaminant Hydrology 119(1-4), 80-88.

Lynch, J.P., Fishbein, M. and Echavarria, M. (2011) Adenovirus. Seminars in Respiratory and Critical Care Medicine 32(4), 494-511.

Masters, C.I., Shallcross, J.A. and Mackey, B.M. (1994) Effect of stress treatments on the detection of listeria-monocytogenes and enterotoxigenic escherichia-coli by the polymerase chain-reaction. Journal of Applied Bacteriology 77(1), 73-79.

Michels, R. and Landais, P. (1994) ARTIFICIAL COALIFICATION - COMPARISON OF CONFINED PYROLYSIS AND HYDROUS PYROLYSIS. Fuel 73(11), 1691-1696.

Muhammad, N., Sinha, R., Krishnan, E.R., Piao, H., Patterson, C.L., Cotruvo, J., Cumberland, S.L., Nero, V.P. and Delandra, C. (2008) Evaluating surrogates for Cryptosporidium removal in point-of-use systems. Journal American Water Works Association 100(12), 98-107.

Murphy, E.M. and Ginn, T.R. (2000) Modeling microbial processes in porous media. Hydrogeology Journal 8(1), 142-158.

Neu, T.R. (1996) Significance of bacterial surface-active compounds in interaction of bacteria with interfaces. Microbiological Reviews 60(1), 151-+.

Okello, J.B.A., Rodriguez, L., Poinar, D., Bos, K., Okwi, A.L., Bimenya, G.S., Sewankambo, N.K., Henry, K.R., Kuch, M. and Poinar, H.N. (2010) Quantitative Assessment of the Sensitivity of Various Commercial Reverse Transcriptases Based on Armored HIV RNA. Plos One 5(11).

Parashar, U.D., Bresee, J.S., Gentsch, J.R. and Glass, R.I. (1998) Rotavirus. Emerging Infectious Diseases 4(4), 561-570.

Parashar, U.D., Burton, A., Lanata, C., Boschi-Pinto, C., Shibuya, K., Steele, D., Birmingham, M. and Glass, R.I. (2009) Global Mortality Associated with Rotavirus Disease among Children in 2004. Journal of Infectious Diseases 200, S9-S15.

Paton, A.W. and Paton, J.C. (1998) Detection and characterization of shiga toxigenic Escherichia coli by using multiplex PCR assays for stx(1), stx(2), eaeA, enterohemorrhagic E-coli hlyA, rfb(O111), and rfb(O157). Journal of Clinical Microbiology 36(2), 598-602.

Peterson, A.A., Vogel, F., Lachance, R.P., Froling, M., Antal, M.J. and Tester, J.W. (2008) Thermochemical biofuel production in hydrothermal media: A review of sub- and supercritical water technologies. Energy & Environmental Science 1(1), 32-65.

Puig, M., Jofre, J., Lucena, F., Allard, A., Wadell, G. and Girones, R. (1994) DETECTION OF ADENOVIRUSES AND ENTEROVIRUSES IN POLLUTED WATERS BY NESTED PCR AMPLIFICATION. Applied and Environmental Microbiology 60(8), 2963-2970.

Redman, J.A., Walker, S.L. and Elimelech, M. (2004) Bacterial adhesion and transport in porous media: Role of the secondary energy minimum. Environmental Science & Technology 38(6), 1777-1785.

Ross, D.S., Loo, B.H., Tse, D.S. and Hirschon, A.S. (1991) HYDROTHERMAL TREATMENT AND THE OXYGEN FUNCTIONALITIES IN WYODAK COAL. Fuel 70(3), 289-295.

Roth, M.J., Tanese, N. and Goff, S.P. (1985) Purification and characterization of murine retroviral reverse transcriptase expressed in Escherichia coli. J Biol Chem 260(16), 9326-9335.

Ryan, J.N. and Elimelech, M. (1996) Colloid mobilization and transport in groundwater. Colloids and Surfaces A: Physicochemical and Engineering Aspects 107(0), 1-56.

Ryan, J.N., Elimelech, M., Ard, R.A., Harvey, R.W. and Johnson, P.R. (1999) Bacteriophage PRD1 and silica colloid transport and recovery in an iron oxide-coated sand aquifer. Environmental Science & Technology 33(1), 63-73.

Schijven, J.F. and Hassanizadeh, S.M. (2000) Removal of viruses by soil passage: Overview of modeling, processes, and parameters. Critical Reviews in Environmental Science and Technology 30(1), 49-127.

Schijven, J.F. and Hassanizadeh, S.M. (2002) Virus removal by soil passage at field scale and groundwater protection of sandy aquifers. Water Science and Technology 46(3), 123-129.

Schuhmacher, J.P., Huntjens, F.J. and Van Krevelen, D.W. (1960) Chemical structure and properties of coal XXVI-studies on artificial coalification. Fuel 39(3), 223-234.

Sevilla, M. and Fuertes, A.B. (2009) Chemical and Structural Properties of Carbonaceous Products Obtained by Hydrothermal Carbonization of Saccharides. Chemistry-a European Journal 15(16), 4195-4203.

Sinton, L.W., Mackenzie, M.L., Karki, N., Dann, R.L., Pang, L. and Close, M.E. (2012) Transport of Escherichia coli and F-RNA Bacteriophages in a 5-M Column of Saturated, Heterogeneous Gravel. Water Air and Soil Pollution 223(5), 2347-2360.

Siskin, M. and Katritzky, A.R. (2001) Reactivity of organic compounds in superheated water: General background. Chemical Reviews 101(4), 825-835.

Stewart, P.L., Fuller, S.D. and Burnett, R.M. (1993) Difference imaging of adenovirus - bridging the resolution gap between x-ray crystallography and electron-microscopy. Embo Journal 12(7), 2589-2599.

Sugimoto, Y. and Miki, Y. (1997) Chemical structure of artificial coals obtained from cellulose, wood and peat, Germany.

Thurston-Enriquez, J.A., Haas, C.N., Jacangelo, J., Riley, K. and Gerba, C.P. (2003) Inactivation of feline calicivirus and adenovirus type 40 by UV radiation. Applied and Environmental Microbiology 69(1), 577-582.

Tipping, E. and Cooke, D. (1982) The effects of adsorbed humic substances on the surface charge of goethite (α-FeOOH) in freshwaters. Geochimica et Cosmochimica Acta 46(1), 75-80.

Titirici, M.M., Thomas, A., Yu, S.-H., Mueller, J.O. and Antonietti, M. (2007) A direct synthesis of mesoporous carbons with bicontinuous pore morphology from crude plant material by hydrothermal carbonization. Chemistry of Materials 19(17), 4205-4212.

Truesdail, S.E., Lukasik, J., Farrah, S.R., Shah, D.O. and Dickinson, R.B. (1998) Analysis of Bacterial Deposition on Metal (Hydr)oxide-Coated Sand Filter Media. Journal of Colloid and Interface Science 203(2), 369-378.

Tufenkji, N., Miller, G.F., Ryan, J.N., Harvey, R.W. and Elimelech, M. (2004) Transport of Cryptosporidium Oocysts in Porous Media: Role of Straining and Physicochemical Filtration†. Environmental Science & Technology 38(22), 5932-5938.

van Halem, D., van der Laan, H., Heijman, S.G.J., van Dijk, J.C. and Amy, G.L. (2009) Assessing the sustainability of the silver-impregnated ceramic pot filter for low-cost household drinking water treatment. Physics and Chemistry of the Earth 34(1-2), 36-42.

Vigeant, M.A.S. and Ford, R.M. (1997) Interactions between motile Escherichia coli and glass in media with various ionic strengths, as observed with a three-dimensional-tracking microscope. Applied and Environmental Microbiology 63(9), 3474-3479.

Vigeant, M.A.S., Ford, R.M., Wagner, M. and Tamm, L.K. (2002) Reversible and irreversible adhesion of motile Escherichia coli cells analyzed by total internal reflection aqueous fluorescence microscopy. Applied and Environmental Microbiology 68(6), 2794-2801.

Vollet, J.J., 3rd, DuPont, H.L. and Pickering, L.K. (1981) Nonenteric sources of rotavirus in acute diarrhea. The Journal of infectious diseases 144(5), 495.

Wait, D.A. and Sobsey, M.D. (1983) Method for recovery of enteric viruses from estuarine sediments with chaotropic agents. Applied and Environmental Microbiology 46(2), 379-385.

Walker, S.L., Redman, J.A. and Elimelech, M. (2004) Role of cell surface lipopolysaccharides in escherichia coli K12 adhesion and transport. Langmuir 20(18), 7736-7746.

WHO (2008) Guidelines for drinking-water quality, third edition, incorporating first and second addenda.

Wong, T.-P., Byappanahalli, M., Yoneyama, B. and Ray, C. (2008) An evaluation of the mobility of pathogen indicators, Escherichia coli and bacteriophage MS-2, in a highly weathered tropical soil under unsaturated conditions. Journal of Water and Health 6(1), 131-140.

Yao, C., Shin, Y., Wang, L.Q., Windisch, C.F., Samuels, W.D., Arey, B.W., Wang, C., Risen, W.M. and Exarhos, G.J. (2007) Hydrothermal dehydration of aqueous fructose solutions in a closed system. Journal of Physical Chemistry C 111(42), 15141-15145.

Zheng, B.J., Chang, R.X., Ma, G.Z., Xie, J.M., Liu, Q., Liang, X.R. and Ng, M.H. (1991) ROTAVIRUS INFECTION OF THE OROPHARYNX AND RESPIRATORY-TRACT IN YOUNG-CHILDREN. Journal of Medical Virology 34(1), 29-37.

Chapter 3: Development of low cost two-step reverse transcription-quantitative polymerase chain reaction assays for rotavirus detection in foul surface water drains

This chapter is based on:

Chung, J.W., Foppen, J.W. and Lens, P.N.L. (2013) Development of low cost two-step reverse transcription-quantitative polymerase chain reaction assays for rotavirus detection in foul surface water drains. Food and environmental virology 5(2), 126-133.

Abstract

Commercial kits to determine RNA concentrations are expensive, and sometimes too expensive for laboratories working with tight budgets, especially those in developing countries. We developed, tested, and evaluated two home-made two-step reverse transcription-quantitative polymerase chain reaction assays aimed to detect rotavirus in surface water samples. A commercial one-step master kit was used for comparison. Our results indicated that the efficiency of the home-made assays was comparable to the commercial kit. Furthermore, the lowest detection limit of all assays corresponded to 10-0.2 TCID50 (50 % Tissue Culture Infective Dose) per ml. The home-made assays were able to detect rotavirus concentrations in complex surface waters in a slum area in Kampala (Uganda) and their performance was comparable to the commercial kit. The total costs of the two home-made assays was 11 times less than the selected commercial kit. Although preparing home-made assays is more time consuming, the assays can be useful for cases in which consumable costs are more important than personnel costs.

3.1 Introduction

Diarrhoeal diseases cause approximately 1.8 million deaths and 4 billion cases of illness every year. The majority of diarrhoeal deaths is among children under 5 year, mostly in developing countries (WHO, 2007). Among those diarrhoeal diseases, acute gastroenteritis caused by rotavirus infection accounts for approximately 527,000 deaths every year (Parashar et al., 2009). Rotavirus is transmitted through the fecal-oral route and possibly also by ingestion of fecally contaminated food and water. Quantitative investigation of rotavirus in environmental water samples is therefore an important issue for examining and ensuring public health in less developed communities. Rotavirus detection can be carried out by several technologies, such as electron microscopy, enzyme-linked immunosorbent assay (ELISA), flow cytometry and reverse transcription quantitative polymerase chain reaction (RT-qPCR). Currently, RT-qPCR has been reported as a sensitive and accurate technique for rotavirus detection (He et al., 2011; Yang et al., 2011).

RT-qPCR can be conducted in two ways: using commercial master kits or using home-made recipes with combinations of reagents. There are many commercial RT-qPCR master kits available on the market, each with their own unique features. Most of these kits are aimed at minimizing time-consuming operational procedures and maximizing ease of use. Also the assurance of quality control encourages the use of commercial kits. Home-made assays are relatively vulnerable to user errors, since more complex manual operations with individual enzymes and chemicals are involved. In contrast to the commercial kits, standardization of home-made assays may be challenging and time-consuming. However, despite the clear advantages of commercial kits, their high costs can be a practical limitation, either when budgets are tight or when large numbers of samples need to be analyzed, for instance in the case of microbial contamination monitoring. Because optimized home-made assays can provide economic advantages, home-made recipes are widely accepted and used in research laboratories and dedicated diagnostic laboratories.

The objective of this paper was to develop, test, and evaluate low cost home-made two-step RT-qPCR assays for the detection of rotavirus. In order to determine the versatility of the home-made assays, both laboratory prepared virus samples and complex environmental samples collected from surface water in a slum in Kampala (Uganda), called Bwaise-III, were tested, whereby we used a commercial one-step kit as a reference.

3.2 Methods and materials
3.2.1 Primers and probe

Primers and probe were taken from Pang et al. (2004), and were targeting the Non-Structural Protein 3 (NSP3) region of group A rotavirus RNA. In our case, the probe was labeled with a 6-carboxyfluorescein (FAM) fluorophore on one side and a Black Hole

Quencher (BHQ) on the other side. The probe and primers were synthesized by Biolegio (Nijmegen, The Netherlands).

3.2.2 Standard virus stock for regression curves

Serial dilutions of a stock of rotavirus Wa strain with a concentration of $10^{5.8}$ TCID$_{50}$ / ml, kindly provided by the National Institute for Health and Environment in the Netherlands (RIVM, Bilthoven, The Netherlands), were used for determining standard curves using the commercial kit and the home-made assays.

3.2.3 Environmental virus samples

To test the versatility of the home-made recipe, six wild-type virus concentrates (named as A5, C3, NS2, P1b, P3a and P3c) obtained from previous research were used. Virions in these samples were isolated from surface water and grey water containing complex biological and chemical components from the Bwaise-III slum, a typical unsewered slum area in Kampala, Uganda (unpublished data). Therefore, virus concentrates were prepared from 10 L water samples with the glass-wool concentration method described by Wyn-Jones et al. (2011), which is a modification from Vilaginès et al. (1993).

3.2.4 Nucleic acid extraction

Viral nucleic acids (NA) of standard virus stock and environmental samples were extracted using a method previously described by Boom et al. (1999). Briefly, 150 µl of sample was treated by 1 mL L7 buffer (5.25 M guanidine thiocyanate (GuSCN), 50 mM Tris-HCl-pH 6.4, 20 mM EDTA, 1.3% [w/v], Triton X-100, and alpha-casein to a final concentration of 1 mg/ml) containing 15 µl silica colloids. The silica-NA complexes were washed twice with 1 ml L2 buffer (5.25 M GuSCN and 50 mM Tris-HCl-pH 6.4), twice with 1 ml 70% ethanol and once with 1 ml acetone. Finally, the nucleic acids were eluted in 80 µl of Tris-EDTA buffer (TE, Fluka).

3.2.5 Commercial one-step RT-qPCR assay

In order to evaluate the home-made assays, a commercial one-step RT-qPCR master kit, the Thermoscript™ one step system (Invitrogen, USA) was selected for comparison. The reaction mixture with a total volume of 25 µl consisted of 12.5 µl 2x ThermoScript™ Reaction Mix, 0.5 µl ThermoScript™ Taq Mix, 6.9 µl DEPC treated water (MP Biomedicals), 0.4 µl forward and reverse primer (200nM final concentration), 0.3 µl probe (150 nM final concentration), and 4 µl sample. The RT-qPCR protocol was 60 °C for 30 min, 95 °C for 5 min, and then 45 cycles of denaturation at 94 °C for 20 s and annealing / extension at 60 °C for 60 s. Fluorescence emission was measured at the end of each extension step. All RT-qPCR work in this research was carried out on a BioRad MJMini™ (real time PCR system, Miniopticon).

3.2.6   Home-made two-step RT-qPCR assay with M-MLV

The RT step in one home-made two-step assay was developed using cheap M-MLV reverse transcriptase (RTase; Promega), random hexamer primer (RH; Fermentas) and RNAse inhibitor (RI; Fermentas). Prior to reverse transcription, denaturation of double stranded RNA of the rotavirus and annealing of RH was carried out. Thereto, a reaction mix containing 2 µl NA extract, 1 µl RH (0.2 µg) and 11 µl DEPC treated water was incubated at 70°C for 5 min. After incubation, the reaction product was immediately chilled on ice, and centrifuged at low speed for a few seconds (Galaxy MiniStar, VWR). Then, cDNA synthesis was carried out. Thereto, to the chilled product, 1 µl (200 unit) M-MLV RTase, 1 µl dNTP mix (4 mM of each dNTP, GenScript), 5 µl 5x RT buffer and 4 µl DEPC treated water were added. We used two 5X RT buffers: one with and one without dithiothreitol (DTT) consisting of 250 mM Tris-HCl at pH 8.3 at 25°C, 15 mM $MgCl_2$, 375 mM KCl and 50 mM DTT, provided by the manufacturer, and a home-made 5x RT buffer of exactly the same composition but without DTT. RT was carried out at 37°C for 60 min and at 70°C for 10 min in the thermal cycler. After the RT reaction, the product, with a total volume of 25 µl, was immediately chilled on ice. This cDNA was used for rotavirus concentration measurements and stored at -80 °C.

The quantitative PCR step for determining cDNA concentrations consisted of adding 4 µl cDNA to a qPCR mix containing 2.5 µl home-made qPCR buffer (150 mM Tris-HCl pH 8.2 at 25°C, 300 mM KCl, 50 mM $(NH_4)_2SO_4$, 25 mM $MgCl_2$, and 0.2% BSA), 1 µl dNTP mix, 1 µl (0.5 unit) Taq polymerase (GenScript), 15.4 µl DEPC treated water, 0.4 µl of forward and reverse primer (both at a 200nM final concentration), and 0.3 µl probe (150 nM final concentration). Then, samples were exposed to a thermal cycling regime consisting of 95 °C for 5 min, and then 40 cycles of denaturation at 94 °C for 20 s followed by annealing / extension at 60 °C for 60 s. The fluorescence signal was measured at the end of each extension step.

3.2.7   Reagent dose optimization in the two-step home-made assay with M-MLV

All standard curves were prepared with reagent doses as given by the manufacturer. Doses of RT, RH and RI were reduced in successive experiments until significantly retarded amplifications were observed. Every reagent dose combination was applied to both high concentration rotavirus Wa samples and the six environmental samples. Mean C_t value and standard deviation of three independent measurements were compared to assess the performance of each reaction specific dose.

3.2.8   Home-made two-step RT-qPCR assay with RevertAid

The RT step in the second home-made assay was developed using RevertAid RTase (Fermentas), whereas the qPCR step was similar to the one described above. Denaturation of double stranded RNA and annealing of random hexamers was carried out by adding 2 µl NA extract, 0.3 µl RH and 11.7 µl DEPC treated water, followed by incubation at 70°C for 5 min. After incubation, the reaction product was immediately chilled on ice and centrifuged at low speed for a few seconds. Then, cDNA synthesis was carried out. Thereto, to the chilled product, 0.3 µl (60 unit) RevertAid RTase, 1 µl dNTP mix, 5 µl 5x RT buffer (250 mM Tris-HCl-pH 8.3 at 25 °C, 20 mM MgCl$_2$, 250 mM KCl and 50 mM DTT) and 4.7 µl DEPC treated water was added. After centrifuging the sample with a final volume of 25 µl for a few seconds at low speed, cDNA synthesis was carried out at 25 °C for 10 min, then at 42 °C for 60 min and finally at 70 °C for 10 min in a thermal cycler. The final product was immediately chilled on ice. This cDNA was used for rotavirus concentration measurements and stored at -80 °C.

3.3 Results

3.3.1 DTT in the RT buffer used in the home-made assay with M-MLV

In the early stages of developing this assay, various combinations of RH, RTase and RI doses did not yield any amplification of the target template when DTT was present in the RT step at a concentration of 10mM, as prescribed by the manufacturer. Additional experiments using newly purchased commercial buffer with different production serial number also did not show any amplification. When an identical home-made RT buffer was prepared without DTT, samples were amplified during the qPCR reaction, and amplification plots had their well known sigmoidal shape. We concluded that apparently DTT was inhibiting either the RT or qPCR step, and we continued using the home-made buffer without DTT.

3.3.2 Standard curves from rotavirus Wa

Using 10-fold dilutions of the rotavirus Wa stock solution, standard curves were prepared with the commercial kit, and with the M-MLV based home-made assay (Figure 3.1). The standard curves were clearly log-linear with R^2-values in excess of 0.99 for seven concentration values ranging between $10^{-0.2}$ and $10^{5.8}$ TCID$_{50}$/ml. The lowest detection limit of the assays was around $10^{-0.2}$ TCID$_{50}$/ml. The amplification efficiency of the commercial kit determined from the standard curve was 125% (Table 3.1), which we considered to be very high. Efficiency of the home-made M-MLV based assay was 101%, and this was considered to be good.

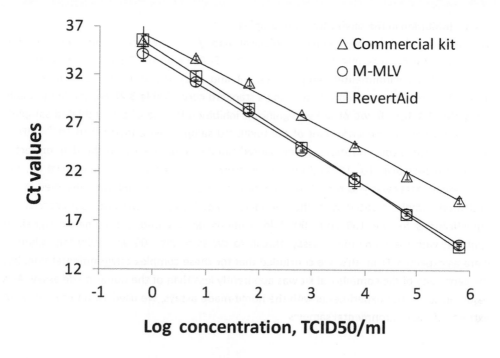

Log concentration, TCID50/ml

Figure 3.1: Standard curves of *Rotavirus Wa* nucleic acid prepared with the commercial kit and with the two two-step home-made assays. Data points and error bars indicate the mean C_t and standard deviation, respectively

Table 3.1: Characteristic values of the standard curves of the assays prepared using the *Rotavirus Wa* strain dilutions. Consumables used were according to the prescription of the manufacturer

Assay		Slope	Intercept	R^2	Amplification efficiency, E(%) [a]
Commercial kit		-2.84	35.68	0.99	125
Home-made	M-MLV	-3.30	33.81	1.00	101
Home-made	RevertAid	-3.55	34.84	1.00	91

[a] Calculated with E = (10^(-1/slope)-1)*100

3.3.3 Inhibition in the environmental samples

We anticipated that the complex biological and / or chemical components in the environmental samples might inhibit the RT, the PCR, or both. Thereto, we diluted NA extracts of 3 environmental samples (A5, C3 and P3c) 10, 100 and 1000 times, and determined rotavirus concentrations using a standard curve (Table 3.2). For the home-made assay (M-MLV based), we observed significant inhibition in case of the undiluted samples: while the rotavirus concentrations of the undiluted samples were lower than $10^{0.6}$ TCID$_{50}$/ ml, the rotavirus concentrations of the diluted samples multiplied with the dilution factor were all higher than 10^2 TCID$_{50}$/ ml. Furthermore, compared to the commercial kit, the home-made assay was less robust for inhibition, as there was no inhibition observed in the undiluted samples treated with the commercial kit. However, with the commercial kit, amplifications for the 100 and 1000-fold dilutions disappeared, rendering no signal. In contrast, with the home-made assay, threshold cycles of the 100 and 1000-fold dilutions were as expected. From this, we concluded that for these complex environmental samples, the sensitivity of the commercial kit was apparently less than of the home-made assay. As a result, in successive experiments with the home-made assays, we used 10 times diluted NA extracts of the environmental samples.

Table 3.2: Comparison of the commercial kit and the home-made M-MLV based assay for a number of diluted environmental samples

	Commercial kit								Home-made assay (M-MLV based)							
Dilutions	1		10		100		1000		1		10		100		1000	
Samples	C_t	Conc.	C_t	Conc	C_t	Conc	C_t	Conc	C_t	Conc	C_t	Conc	C_t	Conc	C_t	Conc
A5	28.4	2.6 [a]	30.3	2.9	n.d.[b]	-	n.d.	-	33.0	0.3	28.1	2.7	31.7	2.7	34.1	2.9
C3	29.4	2.2	33.5	1.8	n.d.	-	n.d.	-	34.3	-0.2	30.0	2.2	33.6	2.1	35.7	2.4
P3c	28.1	2.7	31.1	2.6	n.d.	-	n.d.	-	33.2	0.5	29.2	2.4	32.9	2.3	35.0	2.7

[a] Concentration expressed in log $TCID_{50}$/ ml; concentrations are given for the undiluted samples

[b] n.d.: not detected

3.3.4   Optimization of the reagent dose

The application of RI in the home-made M-MLV based assay was found to be unimportant in both the standard samples and the environmental samples. The omission of RI did not show any consistent trend of retarded C_t values. Furthermore, results with the optimized RH and RTase dose (0.3 µl / reaction) were comparable to results obtained when using the manufacturer's recommended dose (1 µl / reaction). Even the samples with the highest rotavirus concentration were successfully analyzed at optimized low doses of reagents (Table 3.3).

Table 3.3: Reagent dose optimization of the home-made two-step M-MLV based assay

Random hexamer	1 [a]			0.5		0.3		
Reverse transcriptase	1 [a]	0.5	0.3	0.5	0.3	0.3	0.1	
RNase inhibitor	0.5 [a]	0	0	0	0	0	0	
A5	27.9±0.4 [b]	-0.5±0.3 [c]	-0.5±0.5	-0.3±0.2	-0.5±0.0	-0.2±1.0	-0.5±0.7	-5.5±1.5
C3	29.6±0.2	-1.0±0.4	0.1±0.2	0.0±0.3	-0.2±0.3	1.1±0.3	0.5±0.6	-3.3±0.6
NS2	29.0±0.3	-1.3±0.9	0.0±0.6	-0.3±0.4	-0.9±0.1	-0.4±1.2	0.0±0.5	n.d [f]
P1b	29.2±0.4	-0.7±0.8	0.1±0.3	0.1±0.2	-0.9±0.2	0.1±0.2	0.1±0.5	n.d
P3a	28.7±0.2	-1.1±0.2	0.2±0.3	0.4±0.2	-0.3±0.2	0.5±0.1	0.6±0.6	n.d
P3c	28.9±0.2	-0.4±0.3	0.3±0.1	0.2±0.4	0.0±0.0	0.8±0.1	0.3±0.5	n.d
P.C.1 [d]	14.7±0.1	-0.1±0.0	0.2±0.2		0.1±0.1		0.4±0.1	n.d
P.C.2 [e]	17.9±0.2		-0.5±0.4	0.1±0.1		-0.2±0.1	-0.1±0.1	-2.8±0.5

[a] Dose of reagent, µl / 25µl RT reaction mix

[b] Results of the home-made assay for recommended reagent doses prescribed by the manufacturers, C_t ± standard deviation of triplicated measurements

[c] Results of reduced reagent doses, mean C_t subtracted from the results of recommended doses ± standard deviation of triplicated measurements

[d] Positive control: standard sample with a concentration of $10^{5.8}$ $TCID_{50}$/ml

[e] Positive control: standard sample with a concentration of $10^{4.8}$ $TCID_{50}$/ml

[f] n.d. not detected

3.3.5   Performance of the home-made M-MLV based and RevertAid based assays

Encouraged by the results obtained with the home-made M-MLV based assay, we also started to use another cheap reverse transcriptase, RevertAid, using the same optimized doses for RH and RTase, while leaving out RI. For the RevertAid based home-made assay, the efficiency of the standard curve was 91% (Figure. 3.1), and less than the M-MLV based home-made assay. However, we still considered this efficiency to be good.

We compared the performance of the two assays with the performance of the commercial kit for the six environmental samples (Table 3.4). Although only a limited number of environmental samples were tested, we concluded that the results of both home-made assays were comparable to the results obtained with the commercial kit. The largest difference of the mean result between the commercial kit and the home-made assay was $10^{0.29}$ $TCID_{50}$/ ml for the M-MLV based assay and $10^{0.42}$ $TCID_{50}$/ ml for the RevertAid based assay.

Table 3.4: Comparison of the results obtained with the commercial kit and the home-made assays with optimized reagent doses

Samples	Commercial kit	Home-made assay M-MLV based	Home-made assay RevertAid based
A5 [a]	2.55 ± 0.120 [b]	2.63 ± 0.198	2.91 ± 0.059
C3	2.22 ± 0.087	2.43 ± 0.170	2.49 ± 0.107
NS2	2.48 ± 0.310	2.46 ± 0.154	2.35 ± 0.168
P1b	2.73 ± 0.029	2.44 ± 0.162	2.31 ± 0.149
P3a	2.56 ± 0.056	2.72 ± 0.182	2.55 ± 0.253
P3c	2.68 ± 0.096	2.56 ± 0.146	2.73 ± 0.153

[a] Test environmental samples collected from grey water drainage and surface water originating from Bwaise-III slum in Kampala

[b] Mean concentration ± standard deviation of triplicate measurements, determined from the standard curve, expressed in $TCID_{50}$/ml

3.4   Discussion

The application of RI, commonly recommended to be present during cDNA synthesis, did not show any major difference in both standard and environmental samples. We think that this can be explained by the inactivation of RNAses in the NA extraction process using 5.25M GuSCN, which is a high concentration chaotropic salt solution. Also, Bengtsson et al. (2008), and Wolin et al. (1995) reported that all RNAses were completely inactivated in a solution containing 1M GuSCN.

The majority of RT buffers available on the market contain DTT at a final concentration of 10mM in the cDNA synthesis reaction mixture. Otherwise, it is supplied in a separate vial. However, for the home-made two-step M-MLV based assay, the 5x RT buffer provided with the M-MLV RTase did not produce any measureable C_t-value in both standard and environmental samples. Apparently, this lack of amplification was due to the presence of DTT (at a final concentration of 10 mM). The novel buffer prepared in the laboratory with the same composition as the commercial buffer except for DTT enabled successful quantification of rotavirus in all successive experiments and analysis. With regard to the role of DTT as a stabilizer in the RT reaction, also Deprez (2002) reported DTT to be a strong inhibitor in the synthesis of cDNA. However, the buffer supplied with RevertAid RTase did not affect RT-qPCR results, although it also contained DTT in equal concentrations (10 mM final concentration). From this, we concluded that apparently, the effect of DTT in RT-qPCR is strongly dependent on the target template, the type of RTase, and the presence of other medium constituents, and therefore, the use of DTT should be carefully evaluated in RT-qPCR assay development.

Various chemical and biological substances in environmental samples have been reported to have inhibitory effects in PCR. Inhibitory factors include humic compounds, fulvic acids, acidic plant polysaccharides, non-target DNA, bacterial cells, phenolic compounds, heavy metals, formaldehyde, potassium dichromate and fecal matter (Wilson, 1997). Since PCR technologies can only analyze very small sample volumes with sufficient numbers of target template, large volumes of environmental water samples need to be concentrated, and during this concentration stage, undesirable PCR inhibitors can be co-concentrated together with the virions (Ijzerman et al., 1997). Dilution of sample before carrying out RT-qPCR is regarded as the most simple remediation method effectively dealing with inhibitors. However, dilution has adverse impacts, since the lowest detection limit in an environmental sample is inversely proportional to the dilution factor. Target RNA may not be detected when the concentration in undiluted samples is not high enough to compensate for dilutions. Also, high dilution factors will hamper accurate quantification. In this research, the sensitivity of the home-made M-MLV based assay was enough for accurate analysis of rotavirus in 100-1000 diluted samples. In contrast, the sensitivity of the commercial kit reduced tremendously when the environmental samples were diluted in order to eliminate inhibition, since we were unable to amplify the 100-1000 fold dilutions. It is not clear why the M-MLV based home-made assay reads positive at the samples with 1000-fold dilution, whereas the reference 1-step test which performed better at the raw samples without any dilution was unable to read positive even at 100-fold dilution. A possible explanation for the higher sensitivity of the two-step system for samples with low numbers of templates could be the use of two separate buffer solutions in RT and qPCR. The constitution of the buffer solution is a crucial factor in RT-qPCR. While a one-step system should be carried out in a buffer solution that might not be appropriate for the target nucleic acid, either in RT or in qPCR, two-step assays enabling the use of two different buffers have more potential for optimizing specific templates. A similar phenomenon was reported by Shaw et al. (2007). A parallel evaluation of 10 commercial one-step kits and a two-step assay for clinical Foot-Mouth-Disease diagnosis showed a wide range in sensitivity of the commercial one-step kits. Among 10 commercial one-step kits, only 3 products showed comparable (not superior) analytical sensitivity to the two-step assay, while the worst one-step kit was 1000 times less sensitive. Because this research only tested a limited number of specific environmental samples and laboratory samples, further investigations with a wider panel of samples are necessary for validation of the proposed home-made assays. Preliminary inhibition tests are strongly recommended in advance to the intensive implementation in each case of practical use. Considering the successful analysis of the environmental samples from an African slum area subjected to heavy chemical / microbial contaminations with no further than 10-fold dilution of the samples, other environmental samples with less PCR inhibitors are expected to be analyzed without further treatment.

Besides the inhibition observed when using the home-made assays in combination with the environmental samples, another type of inhibition was observed for the commercial kit in the preparation of the standard curve, since the abnormally high amplification efficiency (125%) is usually seen as an inhibitory effect of the RTase (Suslov and Steindler, 2005). Because RTase itself is also a well recognized PCR inhibitor, hindering Taq polymerase activity (Sellner et al., 1992; Fehlmann et al., 1993; Chumakov, 1994; Liss, 2002), the selection and dose of RTase needs to be carefully considered in any RT-qPCR assay development.

The main goal of this research was to minimize the costs of the RT-qPCR assay for rotavirus detection in environmental samples. The largest share of the cost per analysis in RT-qPCR is in the RTase, so therefore we used two of the cheapest RTases available on the market at the time we started this research (September 2011). The cost per sample of both home-made assays was € 0.44-0.49, while the cost per sample of the commercial kit was € 5.80 (Table 3.5). So, the home-made assays were able to accurately determine rotavirus concentrations in complex environmental samples with competitive sensitivity and reproducibility at less than 10% of the costs of the commercial kit selected in this research. A further cost reduction is possible, if the home-made assays can also be used for other RNA viruses, like Norovirus or hepatitis virus, assuming that the affinity of the RH and RTase for other viral RNAs is similar to that of rotavirus.

The cDNA produced from the home-made assays can further be used to detect other templates in single qPCR set-ups, and perhaps even in multiplex cases (Asano et al., 2010; Kottaridi et al., 2012). To complete the comparison between the commercial kit and the home-made assays, the time required to carry out the RT-qPCR analysis for the commercial kit was at least 1 hour less than for the home-made assays (see Table 3.5). In addition, the home-made assays were found to be more labor intensive: the home-made assays required 5 pipetting actions (excluding the preparation of the master mix) and 3 openings / closings of a qPCR well for a sample (annealing of RH, cDNA synthesis and qPCR), whereas the commercial kit needed only 2 pipetting actions for a sample at the beginning of the reaction.

Table 3.5: Comparison of costs and labor of the commercial kit and the home-made assays

Reagent	Commercial kit	Home-made assay M-MLV based	Home-made assay RevertAid based
Reverse transcriptase		0.351	0.302 [a]
Random hexamer	Not applied	0.104	0.104
Taq polymerase +dNTPs		0.037	0.037
Total cost per sample	5.795	0.492	0.443
Time consumption	2h 30min [b]	3h 35min	3h 45min
Pipetting	2 [c]	5	5
PCR well opening / closing	1 [d]	3	3

[a] Local consumer price (EUROs) at Delft (The Netherlands), including tax and handling charges; calculated on the basis of 1000 analyses
[b] Estimated time requirement for the analysis of 48 samples
[c] Number of pipette operations required per sample; master mix preparation is excluded
[d] Number of times the PCR well needs to be opened/closed per sample

In conclusion, we designed two low cost two-step home-made RT-qPCR assays to detect rotavirus concentrations in complex environmental surface water samples with an efficiency and sensitivity comparable to a commercial kit, which we used as a reference in this research. Both home-made assays were 11 times cheaper than the selected commercial kit. The home-made assays, however, were more labor intensive and required longer reaction times than the commercial kit. We think that the substantial cost reduction can compensate these shortcomings, especially in those laboratory settings, where the costs for the reagents are hampering the laboratory in its functioning more than the costs for human labor.

Acknowledgements
This research has been carried as part of research that was funded by the Korean Church of Brussels, Mangu Jeja Church, Seoul, Korea, and the Netherlands Ministry of Development Cooperation (DGIS) through the UNESCO-IHE Partnership Research Fund. It was carried out in the framework of the research project 'Addressing the Sanitation Crisis in Unsewered Slum Areas of African Mega-cities' (SCUSA).

3.5 References

Asano, K.M., de Souza, S.P., de Barros, I.N., Ayres, G.R., Silva, S.O.S., Richtzenhain, L.J. and Brandão, P.E. (2010) Multiplex semi-nested RT-PCR with exogenous internal control for simultaneous detection of bovine coronavirus and group A rotavirus. Journal of Virological Methods 169(2), 375-379.

Bengtsson, M., Hemberg, M., Rorsman, P. and Stahlberg, A. (2008) Quantification of mRNA in single cells and modelling of RT-qPCR induced noise. Bmc Molecular Biology 9.

Boom, R., Sol, C., Beld, M., Weel, J., Goudsmit, J. and Wertheim-van Dillen, P. (1999) f alpha-casein to silica particles. Journal of Clinical Microbiology 37(3), 615-619.

Chumakov, K.M. (1994) Reverse-transcriptase can inhibit PCR and stimulate primer-dimer formation. Pcr-Methods and Applications 4(1), 62-64.

Deprez, R.H.L., Fijnvandraat, A.C., Ruijter, J.M. and Moorman, A.F.M. (2002) Sensitivity and accuracy of quantitative real-time polymerase chain reaction using SYBR green I depends on cDNA synthesis conditions. Analytical Biochemistry 307(1), 63-69.

Fehlmann, C., Krapf, R. and Solioz, M. (1993) Reverse-transcriptase can block polymerase chain-reaction. Clinical Chemistry 39(2), 368-369.

He, X.Q., Cheng, L., Zhang, D.Y., Xie, X.M., Wang, D.H. and Wang, Z. (2011) One-year monthly survey of rotavirus, astrovirus and norovirus in three sewage treatment plants in Beijing, China and associated health risk assessment. Water Science and Technology 63(1), 191-198.

Ijzerman, M.M., Dahling, D.R. and Fout, G.S. (1997) A method to remove environmental inhibitors prior to the detection of waterborne enteric viruses by reverse transcription-polymerase chain reaction. Journal of Virological Methods 63(1-2), 145-153.

Kottaridi, C., Spathis, A.T., Ntova, C.K., Papaevangelou, V. and Karakitsos, P. (2012) Evaluation of a multiplex real time reverse transcription PCR assay for the detection and quantitation of the most common human rotavirus genotypes. Journal of Virological Methods 180(1-2), 49-53.

Liss, B. (2002) Improved quantitative real-time RT-PCR for expression profiling of individual cells. Nucleic Acids Research 30(17).

Pang, X.L.L., Lee, B., Boroumand, N., Leblanc, B., Preiksaitis, J.K. and Ip, C.C.Y. (2004) Increased detection of rotavirus using a real time reverse transcription-polymerase chain reaction (RT-PCR) assay in stool specimens from children with diarrhea. Journal of Medical Virology 72(3), 496-501.

Parashar, U.D., Burton, A., Lanata, C., Boschi-Pinto, C., Shibuya, K., Steele, D., Birmingham, M. and Glass, R.I. (2009) Global Mortality Associated with Rotavirus Disease among Children in 2004. Journal of Infectious Diseases 200, S9-S15.

Sellner, L.N., Coelen, R.J. and Mackenzie, J.S. (1992) Reverse-transcriptase inhibits taq polymerase activity. Nucleic Acids Research 20(7), 1487-1490.

Shaw, A.E., Reid, S.M., Ebert, K., Hutchings, G.H., Ferris, N.P. and King, D.P. (2007) Implementation of a one-step real-time RT-PCR protocol for diagnosis of foot-and-mouth disease. Journal of Virological Methods 143(1), 81-85.

Suslov, O. and Steindler, D.A. (2005) PCR inhibition by reverse transcriptase leads to an overestimation of amplification efficiency. Nucleic Acids Research 33(20).

Vilagines, P., Sarrette, B., Husson, G. and Vilagines, R. (1993) Glass wool for virus concentration at ambient water pH level Water Science and Technology 27(3-4), 299-306.

WHO (2007) Combating waterborne disease at the household level, WHO Press, Geneva, Switzerland.

Wilson, I.G. (1997) Inhibition and facilitation of nucleic acid amplification. Applied and Environmental Microbiology 63(10), 3741-3751.

Wyn-Jones, A.P., Carducci, A., Cook, N., D'Agostino, M., Divizia, M., Fleischer, J., Gantzer, C., Gawler, A., Girones, R., Holler, C., Husman, A.M.D., Kay, D., Kozyra, I., Lopez-Pila, J., Muscillo, M., Nascimento, M.S., Papageorgiou, G., Rutjes, S., Sellwood, J., Szewzyk, R. and Wyer, M. (2011) Surveillance of adenoviruses and noroviruses in European recreational waters. Water Research 45(3), 1025-1038.

Yang, W., Gu, A.Z., Zeng, S.Y., Li, D., He, M.A. and Shi, H.C. (2011) Development of a combined immunomagnetic separation and quantitative reverse transcription-PCR assay for sensitive detection of infectious rotavirus in water samples. Journal of Microbiological Methods 84(3), 447-453.

Chapter 4: Removal of *Escherichia coli* from Saturated Sand Columns Supplemented with Hydrochar Produced from Maize

This chapter is based on:

Chung, J.W., Foppen, J.W., Izquierdo, M. and Lens, P.N.L. (2014) Removal of *Escherichia coli* from saturated sand columns supplemented with hydrochar produced from maize. Journal of Environmental Quality 43(6), 2096-2103.

Abstract

Despite numerous research works on hydrochar utilization, its application in water treatment for pathogen removal still remains unexplored. In this research, we evaluated the efficiency of hydrochar produced from crop residue of maize for water treatment by determining *Escherchia coli* (*E. coli*) breakthrough from sand columns supplemented with hydrochar. In order to enhance the adsorptive capacity, raw hydrochar was activated by 1M KOH at room temperature. The experiments conducted in a 10 cm sand bed with 1.5% (*w/w*) activated and raw hydrochar supplements, not activated by KOH, showed 93% and 72% of *E. coli* removal efficiencies, respectively. KOH activation not only enhanced the *E. coli* removal, but also increased the strength of the attachment: 96% of the *E. coli* retained in the column with activated hydrochar supplements was shown to be irreversibly attached, while this was only 65% for the raw hydrochar. Scanning electron microscopy / energy-dispersive X-ray spectroscopy (SEM-EDX), fourier transform infrared spectroscopy (FTIR) and zeta-potential analyses suggested that these improvements were mainly due to the development of a well-formed porous surface structure and less negative surface charges on the activated hydrochar.

4.1 Introduction

Hydrothermal carbonization (HTC), also known as wet pyrolysis, is the process of converting biomass into carbonaceous material under wet conditions and relatively low temperature (180 - 250 °C). In contrast to the conventional dry pyrolysis (charcoal production), the HTC technology can produce useful carbonaceous materials from various types of biomass with relatively high yields, without the need for an energy-intensive drying step prior to or during the carbonization process (Libra et al., 2011). The detailed mechanism of carbon conversion during HTC has not been fully elucidated yet. At elevated pressure during the HTC process, the biomass is submerged in subcritical water and undergoes complex chemical reactions, including hydrolysis, dehydration, decarboxylation, polymerization and aromatization (Funke and Ziegler, 2010).

Less energy intensive conversion of biowastes into capable adsorbents can make the HTC technology attractive. A wide range of potential applications of HTC has been reported: soil amendment (Steinbeiss et al., 2009; Rillig et al., 2010), energy carrier (Román et al., 2012), lithium battery (Wang et al., 2012), fuel cell (Tusi et al., 2013) and drug delivery (Tian et al., 2011). Also several studies were carried out on the utilization of hydrochar for the removal of abiotic contaminants from aqueous solutions. Despite the obvious potential of hydrochar in Point-Of-Use (POU) water treatment, only few studies have been carried out in this area. Hydrochar derived from animal wastes (poultry litter and swine solids) showed competitive removal efficiencies for polar / apolar endocrine disrupting chemicals (Sun et al., 2011). For heavy metal species, hydrochar produced from switchgrass was reported to be a capable adsorbent for copper and cadmium, while removal performances were further enhanced by alkali activation (Regmi et al., 2012). In addition, uranium(VI), a radioactive contaminant, was successfully removed by hydrochar produced from switchgrass (Kumar et al., 2011) and pine needles (Zhang et al., 2013).

Compared to the conventional activated carbon products, hydrochar has a relatively limited surface area (absence of microporous structure) that is unfavorable for the removal of chemical contaminants. However, oxygenated surface functional groups of hydrochar were pointed out as a common facilitation factor for the removal of these abiotic contaminants (Titirici et al., 2012). These studies demonstrated the potential of hydrochar as an adsorbent in water treatment processes. To our knowledge, there are no reports assessing the effect of hydrochar on the removal of microorganisms from aqueous solutions. Therefore, the objective of this research was to determine the performance of hydrochar as an adsorbent for bacteria. As a representative for plant biomass, hydrochar derived from maize was tested in water - saturated column setups on the removal efficiency of *Escherichia coli*, a fecal indicator organism, from artificial groundwater (AGW).

4.2 Methods and materials

4.2.1 Bacterial suspension

E. coli strain *UCFL-94* was obtained from previous research (Lutterodt et al., 2009). It was enriched in 50 mL LB broth (Oxoid) for 24 h at 37°C and agitated at 120 rpm on an orbital shaker. This *E. coli* stock was refreshed every week. The *E. coli* feeding suspension for column experiments was prepared by diluting the *E. coli* stock in AGW to a concentration of ~1.0 to 1.3 × 10^6 CFU mL^{-1} and was placed on the lab bench for 24 h, at room temperature, for the adjustment of the bacteria in AGW. In all experiments AGW was prepared by dissolving 526 mg L^{-1} CaCl$_2$.2H$_2$O and 184 mg L^{-1} MgSO$_4$.7H$_2$O in demineralized (DI) water, and buffered with 8.5 mg L^{-1} KH$_2$PO$_4$, 21.75 mg L^{-1} K$_2$HPO$_4$ and 17.7 mg L^{-1} Na$_2$HPO$_4$. The resulting pH of the AGW ranged between 6.6 and 6.8 and the electrical conductivity (EC) value ranged between 980 and 1000 µS cm^{-1}.

4.2.2 Hydrochar

A hydrochar stock was purchased from CarbonSolutions (Kleinmachnow, Germany). Residual plant materials of maize were used as a feedstock for HTC. The raw stock of hydrochar was a thick slurry composed of carbonaceous particles with sizes ranging from colloidal scale to millimeters. The initial pH value of the raw hydrochar stock was acidic pH ~ 4. The raw hydrochar stock was washed several times by centrifugation (3 min, at 2,700×g) and hydrochar particles were re-suspended in DI water until the pH value of the suspension became neutral. The suspension was then stored at 4 °C for the use in further experiments.

In order to enhance *E. coli* removal, hydrochar was activated with KOH (Regmi et al., 2012). Next, 5 g (dry weight) hydrochar slurry was suspended in I L KOH. To assess the effect of the strength of KOH on the *E. coli* removal efficiency, 0.1, 0.5, 1.0, and 2.0 mol L^{-1} KOH solutions were used. The hydrochar-KOH suspension was stirred for 1 h at room temperature and washed in DI water, as described above. After washing, the activated hydrochar was immediately mixed with sand and loaded in columns for carrying out *E. coli* flushing experiments (see below).

4.2.3 Column experiments

Experimental setup

99.1% pure quartz sand (Kristallquartz sand, Dorsilit, Germany) was used as a supporting matrix for the hydrochar adsorbents. The particle size distribution of the quartz-sand was obtained from sieve analysis: 0.53% in size range ≤ 0.425 mm, 32.16% in size range 0.425 - 0.560 mm, 40.99 % in size range 0.560 - 1.1 mm, and 26.38 % in size range 1.1 - 1.6 mm, respectively. The quartz-sand was washed with HCl (5%) to remove impurities, followed by rinsing with DI water until the pH of the water became close to fresh DI water (~ 6.8). Sand and hydrochar were thoroughly mixed and loaded in a borosilicate glass column (Omnifit, Cambridge, U.K.) having a 2.5 cm inner diameter and a bed height of 10 cm. The column was tapped and agitated manually to minimize channeling or air entrapment. The hydrochar-sand mixture loaded in the column was carefully treaded down using glass bar throughout the packing process. Then, the column was connected to a pump (MasterFlex model 77201-60, USA) equipped with a manual rotary valve for feed water selection. Prior to each experiment, the column was washed with DI water overnight to remove residual fines and chemicals. At the beginning of each experiment, AGW was flushed into the column until the EC value in the effluent became close to that of the feed AGW. An upward flow rate of 1 mL min^{-1} (0.2 cm min^{-1}) was applied to all column flushing experiments. In order to test if the column was properly constructed, and to determine pore volume and dispersion, a tracer test was carried out (in triplicate) using a NaCl solution prior to the experiments. Briefly, 45 mL of 20 mmol L^{-1} NaCl solution was flushed into the column with an upward flow rate of 1 mL min^{-1}, followed by 45 mL of DI water flushing. The concentration of Cl$^-$ in the effluent was measured in 3 min intervals by performing ion chromatography (ICS-1000, Dionex, USA).

Assessing the effect of KOH concentration
E. coli removal efficiencies of hydrochar prepared with various KOH concentrations were determined by breakthrough analyses from column experiments. The columns were prepared as described above, using either sand, or sand with 2% (*w/w*) raw or activated hydrochars. For each column experiment, 45 mL of a *E. coli* - AGW suspension was flushed into the column with an upward flow rate of 1 mL min^{-1} (loading phase), followed by 45 mL of *E. coli* free AGW flushing (deloading phase). The effluent samples were collected in 5 min intervals and the concentration of *E. coli* was determined by conventional plate counting method. The effluent samples were spread on selective agar plates (ChromoCult® Coliform Agar, Merck, Germany) in duplicate of 10-fold serial dilutions (APHA, 1998). The plates were incubated at 37°C for 24 h. The colonies were manually enumerated using the Colony Counter (IUL, Barcelona, Spain). Each experimental setup was tested in duplicate.

Assessing the effect of hydrochar concentration on *E. coli* removal
We used adsorbent doses of 1.5% *w/w* raw (not activated with KOH) and 0.5, 1.0, 1.5% (*w/w*) hydrochar supplements activated with KOH to determine *E. coli* removal via breakthrough analyses in the packed bed columns (Table 4.1). For each breakthrough analysis, the loading and deloading of *E. coli* was conducted by flushing a 50 mL of *E. coli* - AGW suspension followed by 50 mL of *E. coli* free AGW flushing with an upward flow rate of 1 mL min^{-1}. The concentration of *E. coli* in the effluent was measured at 5 min intervals by the plate counting technique. The statistical differences of mean removal efficiencies were identified by One-way ANOVA. The mean removal efficiencies were separated by Tukey's honestly significant difference test ($p < 0.05$). All statistical analyses were carried out using SPSS Statistics v. 20 (IBM, 2012)

Table 4.1: Design of breakthrough experiments

Hydrochar	Conc. (%)	Feeding agents in breakthrough analyses		
		50 mL E. coli - AGW + 50 mL AGW	50 mL AGW backwashing	300 mL DI water flushing
Raw	1.5[a]	7[b]	3[c]	2[c]
Activated	1.5	17	5	4
	1.0	7	3	2
	0.5	7	3	2

[a] designed dose of hydrochar supplements in sand media (w/w)
[b] number of breakthough analyses
[c] number of additional tests, either backwashing or DI water flushing

Backwashing

For selected experiments, after the E. coli loading and deloading phases, pumping was stopped, the column was turned upside down, and then, 50 mL of AGW was flushed in upward direction at a rate of 1 mL min[-1] (Table 4.1). The backwashing experiments were conducted assuming that E. coli, which was physically strained in the column was not irreversibly attached to the surface of the media in the column, and therefore, reversing the flow could release a part of the strained E. coli from the column (Foppen et al., 2007a)

DI water flushing

For selected experiments, 300 mL of E. coli free DI water was flushed after the deloading phase. It was assumed that a low ionic strength of DI water increased the double layer repulsion between E. coli and the surface of the media inside the column. Therefore, loosely attached E. coli or those deposited in the secondary energy minimum would be released from the media (Foppen et al., 2007b).

Spatial hydrochar concentration in the column

At the end of each breakthrough test, the packing material in the column was excavated and sliced into 10 equal sections of 1 cm each, in order to determine the vertical hydrochar concentration distribution. Excavated samples were dried (105°C, overnight) and incinerated (500°C, 30 min) subsequently, and the weight of samples were measured at each stage. The weight of the residual adsorbent and sand-adsorbent ratio was calculated from the weight loss resulting from incineration. Reference information for the incinerable proportion of hydrochar was obtained by performing identical tests on fresh hydrochar samples.

4.2.4 Material characterization

Zeta potential - pH titration

The zeta potential values of raw and activated hydrochar and *E. coli* were monitored as a function of pH (4 - 10 for adsorbents and 5 - 9 for *E. coli*) using a Zetasizer Nano ZS (Malvern, UK) equipped with a MPT-2 pH auto-titration unit. The *E. coli* and hydrochar samples for zeta potential measurements were prepared by washing and centrifugation using AGW. All test samples were diluted until an adequate attenuator (6 - 8) was selected by the instrument.

Scanning Electron Microscopy Energy–Dispersive X-Ray Spectroscopy

SEM images of the selected adsorbent surfaces were obtained using an S-4100 scanning electron microscope (Hitachi Ltd., Japan) at 10 kV, equipped with a field emission gun, a BSE Autrata detector, and a Röntec XR detector. Gold coating was used for image capturing processes. EDX spectra was recorded with an XL-30 ESEM (Philips, The Netherlands), equipped with an EDAX microanalysis system. Semi-quantitative elemental composition (relative proportions of elements) of sample was derived from the EDX results.

Fourier Transform Infrared Spectroscopy

FTIR spectra of selected hydrochar particles were recorded using a Thermo Nicolete Nexus FTIR spectrometer (Thermo Fisher Scientifics) equipped with a DTGS / KBr detector and a KBr beam. The KBr disk technique was used for sample preparation, with a mass percentage of the carbon sample of 2.5% and a total mass of each disk of 60 mg. CO_2, aqueous vapor in air, and possible KBr impurity interferences were minimized by collecting a KBr background for each analysis. Spectra were recorded in the 400 - 4000 cm^{-1} range at a 2 cm^{-1} resolution after 128 scans.

4.3 Results

4.3.1 Adsorbent selection

The earliest appearance of *E. coli* in effluent sample was observed from the hydrochar amended columns within 10 min followed the samples from sand columns. *E. coli* may travel faster in hydrochar amended columns because the free space in the sand matrix was occupied by hydrochar supplements resulting in less pore volume. For all experiments, the breakthrough curves (BTCs) showed a clear pattern consisting of a rising limb, followed by a plateau phase and then, finally, a falling limb (Figure 4.1). Furthermore, the plateau C/C_0 values from BTCs in columns either with raw hydrochar or activated hydrochar were lower than values obtained from the quartz sand column only. Prior activation of the hydrochar with a 1.0 and 2.0 mol L^{-1} KOH solution showed an additional increase in *E. coli* removal with a maximum $C/C_0 < 0.05$. In contrast, the plateau C/C_0-values of *E. coli* breakthrough curves in columns with hydrochar which was activated with 0.1, 0.5 mol L^{-1} KOH and without activation (raw hydrochar) were invariably around 0.2. Since there was no further enhancement of the *E. coli* removal efficiency for KOH concentrations above 1 mol L^{-1}, all successive experiments were carried out using hydrochar which was activated prior to packing in the columns with a 1.0 mol L^{-1} KOH solution.

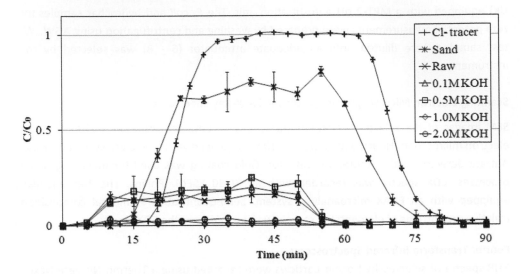

Figure 4.1: *E. coli* breakthrough curves for sand, 2% raw hydrochar, 2% activated hydrochar (0.1, 0.5, 1.0 and 2.0 mol L^{-1} KOH) and chloride breakthrough curves for sand: solid lines represent the mean C/C$_0$ value of duplicated analyses and error bars indicate individual data points.

4.3.2 Material characterization
Zeta potential

The zeta potential values of all adsorbents and *E. coli* strain *UCFL-94* were negative in the pH range of 6.6 - 6.8, which was used for the column experiments (Figure 4.2). Apparently, KOH activation increased the zeta potential of activated hydrochar to a certain extent: while the zeta potential value of raw hydrochar was -10.9 ± 0.84 mV (average ± standard deviation, n=3), activated hydrochar had less negative zeta potential values -7.23 ± 2.02 mV (n=5). Furthermore, the *E. coli* suspension had a negative zeta-potential of around -11.45 ± 1.10 mV (n=4).

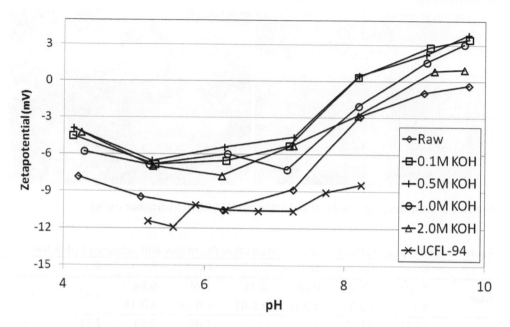

Figure 4.2: Zeta potential of hydrochar adsorbents and *E. coli* as a function of pH in artificial groundwater

Scanning Electron Microscopy Energy–Dispersive X-Ray Spectroscopy

SEM images of hydrochar samples are shown in Figure 4.3. A relatively smooth and less porous surface was found in the raw sample, while the KOH activated sample showed a more rough and porous surface. We also observed that the interaction between the adsorbent surface and *E. coli* increased due to the development of a macroporous surface structure that was easily accessible for *E. coli* having a size of 2 μm in our tests (measured by the Zetasizer Nano ZS, data not shown). Semi-quantitative information of the elemental composition was obtained from EDX analysis (Table 4.2). After KOH treatment, distinctive changes were visible, like the disappearance of phosphorus (P) and the appearance of potassium (K) and aluminium (Al), respectively. These results confirmed the sorption of K to the carbon surface. The presence of Al in activated hydrochar was most probably due to the impurities present in KOH.

Figure 4.3: SEM image of raw (left) and activated (1M KOH) hydrochar (right)

Table 4.2: Semi-quantitative elemental composition (%) of raw and activated hydrochar

	C	O	Si	P	Fe	F	K	Al
Raw	75.03[a] ± 1.16	22.55 ± 1.02	0.48 ± 0.17	0.31 ± 0.02	1.24 ± 0.07	0.38 ± 0.14	n.d.[b]	n.d.
Activated	72.21 ± 1.09	23.54 ± 1.16	0.7 ± 0.08	n.d.	1.46 ± 0.15	0.55 ± 0.20	2.35 ± 0.22	0.20 ± 0.03

[a] average wt % ± standard deviation of triplicated analyses
[b] not detected

Fourier Transform Infrared Spectroscopy

Figure 4.4 illustrates the FTIR spectra of hydrochar adsorbents. The spectra of raw and activated hydrochar samples showed a similar trend of peak developments. The activation with KOH did not significantly alter the surface functional groups of hydrochar. A broad peak at 3342 cm^{-1} was attributed to the O-H stretching vibrations of the aromatic and aliphatic groups. The two peaks at 2923 and 2852 cm^{-1} were due to the asymmetric and symmetric C-H stretching vibrations of aliphatic structures such as CH, CH$_2$ and CH$_3$. The peak occurring at 1699 cm^{-1} was attributed to the C=O stretching of carbonyl, quinone, ester or carboxyl groups (Kumar et al., 2011; Kong et al., 2013). The peaks at 1610 and 1513 cm^{-1} were due to the C=C and / or C=O stretching from aromatic structures, and the C=C stretching in aromatic rings characteristic of lignin, respectively (Zaccheo et al., 2002; Kumar et al., 2011; Kong et al., 2013). Peaks characteristic of cellulose appear in the spectra of carbon, and these peaks correspond to 1161, 1113 and 1061 cm^{-1}, respectively (Figure 4.4). The peak at 1161 cm^{-1} was due to C-O-C bridge asymmetric stretching, while the peak at 1113 cm^{-1} was due to in-phase ring asymmetric stretching, and the peak at 1061 cm^{-1} due to C-O stretch skeletal vibrations (Zaccheo et al., 2002). These results clearly indicated that hydrochar had a cellulose-like molecular structure, and that the maize was not completely decomposed in the hydrothermal reaction (Kong et al., 2013). After the KOH activation, the peaks at 1699 and 1610 cm^{-1} had disappeared, and a broad band at 1592 cm^{-1} in the spectrum of the activated hydrochar sample had appeared. We attributed this to the carboxylate group (Yazdanbakhsh et al., 2009).

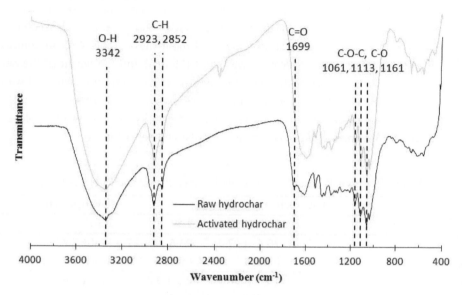

Figure 4.4: FTIR spectra of raw and activated hydrochar

4.3.3 Breakthrough analyses

BTCs for four hydrochar concentrations (0.5, 1.0, 1.5% *w/w*, after activation with 1 mol L^{-1} KOH, and 1.5% raw hydrochar, respectively) were determined, including backwashing and DI water flushing tests to investigate the *E. coli* retention mechanisms in the column. The *E. coli* removal efficiency, the proportion of *E. coli* released due to backwashing or DI flushing, and the amount of residual hydrochar supplement in the column media are given in Table 4.3. The natural die-off of *E. coli* in AGW was found to be negligible during the 48 h of monitoring (data not shown). A high average *E. coli* removal efficiency (93%) was obtained from 1.5% supplements of activated hydrochar. The results from 0.5%, 1.0% activated and 1.5% raw hydrochar showed similar *E. coli* removal performances (75%). Although there were fractional differences in the mean *E. coli* removal efficiencies of these three types of supplements, the relatively large standard deviations indicated a lack of significance. In comparison to the results obtained from 0.5% and 1.0% activated hydrochar supplements, we observed that a 2-fold increase in the amount of hydrochar in the columns did not significantly improve the *E. coli* removal in the column. In this range, between 0.5 and 1.0%, the amount of activated hydrochar was apparently not a limiting factor for the *E. coli* removal.

In order to understand the mechanism of *E. coli* removal in the columns, backwashing and DI water flushing tests were carried out after the deloading phase with *E. coli* free AGW.

With the backwashing tests, *E. coli* was released from the column (Figure 4.5). Release from raw hydrochar supplements was substantial with a high peak ($C/C_0 = 1.4$) within 5 min after commencing the backwash test, and then, for the next 10 min, a declining peak, followed by a gentle release for the last 15 min of the test. The total amount of *E. coli* released during the backwashing step was around 21% of the total input mass (Table 4.3). In contrast, release of *E. coli* from activated hydrochar was short (15 min) and the height of the peak was low ($C/C_0 < 0.1$), regardless of the activated hydrochar concentration. When DI water was flushed into the column after the deloading phase, only a minor proportion (< 5%) of the *E. coli* was released (see Figure 4.6 and Table 4.3). At first, small peaks were visible, and later, after 20 min, the release of *E. coli* became irregular and negligible.

Table 4.3: Results from breakthrough analyses with backwashing and DI water flushing

Content	Activated hydrochar			Raw hydrochar
	0.5[a]	1.0	1.5	1.5
Removal efficiency [b]	73.6 ± 10.3 [e] (n=7)	76.4 ± 7.4 [e] (n=7)	93.6 ± 4.2 (n=17)	72.4 ± 7.7 [e] (n=7)
E. coli release in Back washing	1.2 ± 0.9 (n=3)	1.2 ± 0.9 (n=3)	2.1 ± 0.2 (n=5)	21.7 ± 3.4 (n=3)
E. coli release in DI water flushing	0.3 ± 0.2 (n=2)	0.7 ± 0.4 (n=2)	1.5 ± 2.0 (n=5)	3.1 ± 1.6 (n=2)
E. coli remained in column [c]	72.2 (98.0)	74.5 (97.5)	90.0 (96.1)	47.6 (65.8)
Residual hydrochar [d]	0.37 ± 0.04 (n=7)	0.64 ± 0.14 (n=7)	1.01 ± 0.10 (n=17)	1.48 ± 0.14 (n=7)

[a] designed dose of hydrochar supplements % (*w/w*)

[b] average % ± standard deviation or individual data points (if n=2)

[c] an approximate estimation of irreversibly attached *E. coli* in column media: removal efficiency - *E.coli* released due to backwashing - *E.coli* released due to DI water flushing. The number in the parenthesis represent the proportion in total removal

[d] residual adsorbent in the column after breakthrough analyses, average % ± standard deviation (*w/w*)

[e] mean removal efficiencies marked with # are not significantly different using Tukey's honestly significant difference test at $p < 0.05$.

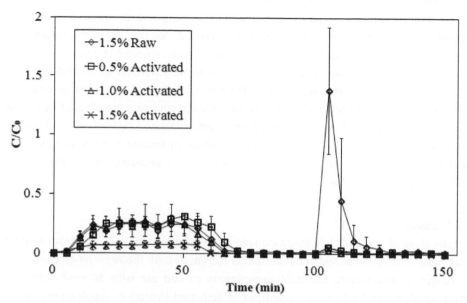

Figure 4.5: Breakthrough curves with additional back flushing: solid lines represent the mean C/C_0 value and error bars indicate the standard deviation. Note that the back flushing started at 100 min.

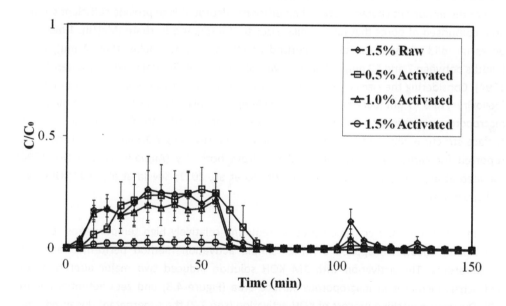

Figure 4.6: Breakthrough curves with additional DI water flushing: solid lines represent the mean C/C_0 value and error bars indicate either standard deviation or individual data points (if n=2). Note that the DI water flushing started at 100 min, and the values later than 150 min was omitted due to insignificant C/C_0.

Based on the information obtained from the backwashing and DI water flushing tests described above, we estimated the irreversibly retained *E. coli* fraction in the column. For the activated hydrochar supplement, > 96% of the *E. coli* removed in the breakthrough analysis was regarded as firmly attached, regardless of the adsorbent dose, while this was only 65% for the raw hydrochar supplement. The concentrations of hydrochar in excavated samples were more or less even along the length of each column, regardless of hydrochar type and dose (data not shown). In column tests with activated hydrochar supplements, approximately 2/3 of the initially added (prior to activation) amount of adsorbents was still present in the excavated packing material, whereas this amounted to > 98% in raw hydrochar supplements (Table 4.3).

4.4   Discussion

This study showed the ability of maize-derived hydrochar to remove *E. coli* from artificial ground water. Activation of hydrochar with KOH increased both the removal efficiency and the strength of attachment. Lab-scale experiments carried out with 10 cm-height sand columns supplemented with minor quantities of activated hydrochar supplements caused more than 90% of *E. coli* to be removed.

4.4.1 Effect of KOH activation on hydrochar

One of the important characteristics of an efficient adsorbent is to provide sufficient surface area comprised of pores that have similar sizes to the target adsorbate (Martin, 1980). The pores of solid material can be categorized as micro (widths smaller than 2 nm), meso (widths between 2 and 50 nm) and macro (widths larger than 50 nm) pores (Rouquerol et al., 1994). Considering the sizes of waterborne pathogenic microorganisms, ranging from 30 nm (poliovirus) to 5 - 7 µm (oocysts of *Cryptosporidium*), it is advantageous to maximize the meso-macroporous structures on the surface of the adsorbent. In order to obtain a highly porous surface structure, different types of activation procedures using KOH have been tested and reported for carbonized materials produced from both dry (Ahmadpour and Do, 1996; Hameed et al., 2007; Li et al., 2012) and wet (Li et al., 2011; Sevilla et al., 2011; Qi et al., 2012) pyrolysis.

In contrast to the KOH activation process requiring relatively high temperatures (Bagheri and Abedi, 2009), we applied a rather simple activation method performed at room temperatures. The activation with 1M KOH solution induced two major alterations on hydrochar: increases in macroporous surface area (Figure 4.3) and zeta-potential (Figure 4.2). During the washing process of KOH activation (see 2.2) the supernatant decanted after the first centrifugation had a somewhat dark brown colour. Subsequently, when the washing process was repeated, the brownish color became much lighter. The source of this colour was presumably due to the dissociation of alkali-soluble fractions present in the raw hydrochar. This can also be supported by the results obtained from the measurements on the residual HTC adsorbent present in the columns. While more than 98% of the hydrochar was present in excavated samples from columns with raw hydrochar supplement, it was only around 2/3 for the columns with activated hydrochar (Table 4.3). Only minimal hydrochar supplements were present in the effluent throughout the column operation, also after column washing and during breakthrough analysis. Therefore, the major cause of weight loss was due to the dissolution of the alkali-soluble fraction of raw hydrochar during KOH activation. Similarly, Regmi et al. (2012) reported that cold-alkali activation of switchgrass-derived hydrochar resulted in enhanced pore structures, while the alteration of surface functional groups remained insignificant. It was apparent that KOH activation cleaned up the pores that were partially blocked with the products from re-polymerization / re-condensation during the HTC process.

Since the FTIR analysis demonstrated only minor alterations of surface functional groups followed KOH activation, the elevated zeta-potential could mainly be attributed to the association of the K^+ ion with the negatively charged surface functional groups of raw hydrochar which might facilitate the attachment of the E. coli on the adsorbent surfaces to a certain extent. In the EDX analysis, the appearance of K in the activated hydrochar demonstrates the deposition of the K^+ ion on the hydrochar surfaces. The disappearance of P can be explained by washing out of partially degraded cellulose under alkaline conditions (Knill and Kennedy, 2003). Despite of low concentration (0.2%), the occurrence of Al would have facilitated the retention of E. coli to a certain extent thanking to strong bridging effect of trivalent cations.

4.4.2 Reversibility of *E. coli* attachment to hydrochar

Transport and retention of E. coli in saturated columns packed with heterogeneous media are governed by complex processes involving the chemistry of the solution, surface properties of E. coli and column media, grain size, and flow rate (Foppen and Schijven, 2006). In order to have a better understanding of the mechanism of E. coli attachment on the adsorbent surfaces, we carried out backwashing and DI water flushing tests. Backwashing tests of columns with both raw and activated hydrochar supplements demonstrated the release of E. coli from the column to a certain extent. It was apparent that the hydrodynamic shear forces during backwashing were strong enough to overcome at least a part of the adhesive forces of the physically strained cells (Foppen et al., 2007a). Since mobilized cells during backwashing may not have been fully released into the effluent due to re-straining and/or re-attachment under these reversed flow conditions, we considered the results as a qualitative and comparative indicator. In that sense, our results clearly demonstrated the difference in E. coli retention mechanism between raw and activated hydrochar supplements, whereby straining played a more significant role in raw hydrochar than in activated hydrochar (Table 4.3).

As DI water flushing decreased the ionic strength in the column, the electrostatic repulsive force between *E. coli* cells and the surface of the adsorbent increased, resulting in the release of *E. coli* cells that were expected to be retained in the secondary energy minimum (Foppen et al., 2007b). Considering the results from the DI water flushing tests, we concluded that the release of *E. coli* in all experiments was low regardless of adsorbent type (1.42 ± 1.73%, mean value ± standard deviation, n=10). This clearly demonstrated that the secondary energy minimum did not play an important role in our experiments. Apparently, the majority of *E. coli* cells attached on the hydrochar seemed to reside in the primary energy minimum. These results from backwashing and DI water flushing tests confirmed that the activation with KOH improved not only the removal efficiency of *E. coli* but also the strength of the attachment, and that these two factors positively enhanced the performance of *E. coli* removal.

Two major alterations induced by the KOH activation, development of the less negative surface charge and macroporous surface structure, enhanced the *E. coli* removal / retention. It was reported that an increase of surface roughness of the collector has a favorable effect on colloid deposition on solid surfaces (Hoek and Agarwal, 2006; Morales et al., 2009; Shen et al., 2012).

The attachment of bacteria on solid surfaces can occur only when the attractive forces, such as van der Waals forces or hydrophobic interactions, exceed the electrostatic repulsion between negatively charged surfaces (Marshall, 1986; Rijnaarts et al., 1995). The increased zeta-potential of the activated hydrochar reduced the repulsion between the surfaces of hydrochar and *E. coli*. The attachment of *E. coli* on negatively charged surfaces could be attributed to heterogeneously distributed surface charges on the adsorbent (Truesdail et al., 1998). Locally, there could be positively or less negatively charged sites on the hydrochar surface, which would facilitate *E. coli* attachment. Thus, it is worth mentioning that the zeta-potential alone might not be a useful parameter for predicting the interaction between microorganisms and heterogeneous surfaces, since it only represents the net or average surface charge (Elimelech et al., 2000).

4.4.3 Suggestions for the further research

Since additional activation processes may not be desirable due to the extra costs and environmental burdens from by-products, the direct synthesis of meso-macroporous hydrochar can be beneficial. The characteristics of hydrochar are usually determined by the nature of the feed stock and several HTC parameters such as hydrous condition, temperature, residence time, pressure, solid load, and pH (Funke and Ziegler, 2010). Accordingly, the development and optimization of an HTC process for meso-macroporous hydrochar is an important topic for further research. As an example of direct synthesis, mesoporous hydrochar was produced from crude plant materials without further activation process (Titirici et al., 2007): pine cones, pine needles and oak leaves were hydrothermally carbonized at 200°C for 16 h in the presence of a minor amount of additive citric acid. The resulting hydrochar showed pore sizes ranging between 10 and 100 nm.

In addition to porous surface structure, hydrophobicity is an important factor determining bacterial sorption to carbonaceous surfaces (Abit et al., 2012; Abit et al., 2014). Quantitative investigations on the hydrophobicity and specific surface area (e.g. Brunauer-Emmett-Teller, BET method) of hydrochar adsorbents would enlarge understanding on KOH treatment and related increase of *E. coli* removal.

The results of this research generally agreed with recent researches on biochar produced from dry pyrolysis of pine chips and poultry litter. In similar experimental conditions of 10 cm bed height with 1-2 % (*w/w*) adsorbent dose, pine chip-based biochar pyrolyzed at 700°C showed comparable bacteria removal efficiency (~98%) to activated hydrochar amendment (Abit et al., 2012; Bolster and Abit, 2012; Abit et al., 2014). A cost comparison between dry pyrolysis and HTC would provide general guidelines for selection of carbonization technology.

Several hydrochars produced from biowastes were reported as capable adsorbents for a wide range of abiotic contaminants in aqueous solutions such as heavy metal species, radioactive agent and polar / apolar organic chemicals (Kumar et al., 2011; Sun et al., 2011; Regmi et al., 2012). Encouraged by the results from our research, an interesting topic for further research is the simultaneous removal of biotic / abiotic contaminants from water using hydrochar-based adsorbent.

Acknowledgements

This research was funded by the Korean Church of Brussels, Mangu Jeja Church, Seoul, Korea, and the Netherlands Ministry of Development Cooperation (DGIS) through the UNESCO-IHE Partnership Research Fund. It was carried out in the framework of the research project 'Addressing the Sanitation Crisis in Unsewered Slum Areas of African Mega-cities' (SCUSA). Authors wish to thank the Instituto de Ciencias de los Materiales of the Universitat de Valencia (ICMUV) for its collaboration with the FTIR analysis.

4.5 References

Abit, S.M., Bolster, C.H., Cai, P. and Walker, S.L. (2012) Influence of Feedstock and Pyrolysis Temperature of Biochar Amendments on Transport of Escherichia coli in Saturated and Unsaturated Soil. Environmental Science & Technology 46(15), 8097-8105.

Abit, S.M., Bolster, C.H., Cantrell, K.B., Flores, J.Q. and Walker, S.L. (2014) Transport of Escherichia coli, Salmonella typhimurium, and Microspheres in Biochar-Amended Soils with Different Textures. Journal of Environmental Quality 43(1), 371-388.

Ahmadpour, A. and Do, D.D. (1996) The preparation of active carbons from coal by chemical and physical activation. Carbon 34(4), 471-479.

APHA (ed) (1998) Standard Methods for the Examination of Water and Wastewater (20th ed.), American Public Health Association, Washington, D.C.

Bagheri, N. and Abedi, J. (2009) Preparation of high surface area activated carbon from corn by chemical activation using potassium hydroxide. Chemical Engineering Research and Design 87(8), 1059-1064.

Bolster, C.H. and Abit, S.M. (2012) Biochar Pyrolyzed at Two Temperatures Affects Escherichia coli Transport through a Sandy Soil. Journal of Environmental Quality 41(1), 124-133.

Elimelech, M., Nagai, M., Ko, C.H. and Ryan, J.N. (2000) Relative insignificance of mineral grain zeta potential to colloid transport in geochemically heterogeneous porous media. Environmental Science & Technology 34(11), 2143-2148.

Foppen, J.W., van Herwerden, M. and Schijven, J. (2007a) Measuring and modelling straining of Escherichia coli in saturated porous media. Journal of Contaminant Hydrology 93(1–4), 236-254.

Foppen, J.W., van Herwerden, M. and Schijven, J. (2007b) Transport of Escherichia coli in saturated porous media: Dual mode deposition and intra-population heterogeneity. Water Research 41(8), 1743-1753.

Foppen, J.W.A. and Schijven, J.F. (2006) Evaluation of data from the literature on the transport and survival of Escherichia coli and thermotolerant coliforms in aquifers under saturated conditions. Water Research 40(3), 401-426.

Funke, A. and Ziegler, F. (2010) Hydrothermal carbonization of biomass: A summary and discussion of chemical mechanisms for process engineering. Biofuels Bioproducts & Biorefining-Biofpr 4(2), 160-177.

Hameed, B.H., Din, A.T.M. and Ahmad, A.L. (2007) Adsorption of methylene blue onto bamboo-based activated carbon: Kinetics and equilibrium studies. Journal of Hazardous Materials 141(3), 819-825.

Hoek, E.M.V. and Agarwal, G.K. (2006) Extended DLVO interactions between spherical particles and rough surfaces. Journal of Colloid and Interface Science 298(1), 50-58.

Knill, C.J. and Kennedy, J.F. (2003) Degradation of cellulose under alkaline conditions. Carbohydrate Polymers 51(3), 281-300.

Kong, L., Miao, P. and Qin, J. (2013) Characteristics and pyrolysis dynamic behaviors of hydrothermally treated micro crystalline cellulose. Journal of Analytical and Applied Pyrolysis 100(0), 67-74.

Kumar, S., Loganathan, V.A., Gupta, R.B. and Barnett, M.O. (2011) An Assessment of U(VI) removal from groundwater using biochar produced from hydrothermal carbonization. Journal of Environmental Management 92(10), 2504-2512.

Li, M., Li, W. and Liu, S.X. (2011) Hydrothermal synthesis, characterization, and KOH activation of carbon spheres from glucose. Carbohydrate Research 346(8), 999-1004.

Li, Z., Zhang, L., Amirkhiz, B.S., Tan, X.H., Xu, Z.W., Wang, H.L., Olsen, B.C., Holt, C.M.B. and Mitlin, D. (2012) Carbonized Chicken Eggshell Membranes with 3D Architectures as High-Performance Electrode Materials for Supercapacitors. Advanced Energy Materials 2(4), 431-437.

Libra, J.A., Ro, K.S., Kammann, C., Funke, A., Berge, N.D., Neubauer, Y., Titirici, M., Fuhner, C., Bens, O., Kern, J. and Emmerich, K. (2011) Hydrothermal carbonization of biomass residuals: A comparative review of the chemistry, processes and applications of wet and dry pyrolysis. Biofuels 2:89-124.

Lutterodt, G., Basnet, M., Foppen, J.W.A. and Uhlenbrook, S. (2009) The effect of surface characteristics on the transport of multiple Escherichia coli isolates in large scale columns of quartz sand. Water Research 43(3), 595-604.

Marshall, K.C. (1986) Adsorption and adhesion processes in microbial growth at interfaces. Advances in Colloid and Interface Science 25(0), 59-86.

Martin, R.J. (1980) Activated Carbon Product Selection for Water and Wastewater Treatment. Industrial & Engineering Chemistry Product Research and Development 19(3), 435-441.

Morales, V.L., Gao, B. and Steenhuis, T.S. (2009) Grain Surface-Roughness Effects on Colloidal Retention in the Vadose Zone. Vadose Zone Journal 8(1), 11-20.

Qi, X.H., Guo, H.X., Li, L.Y. and Smith, R.L. (2012) Acid-Catalyzed Dehydration of Fructose into 5-Hydroxymethylfurfural by Cellulose-Derived Amorphous Carbon. Chemsuschem 5(11), 2215-2220.

Regmi, P., Moscoso, J.L.G., Kumar, S., Cao, X.Y., Mao, J.D. and Schafran, G. (2012) Removal of copper and cadmium from aqueous solution using switchgrass biochar produced via hydrothermal carbonization process. Journal of Environmental Management 109, 61-69.

Rijnaarts, H.H.M., Norde, W., Bouwer, E.J., Lyklema, J. and Zehnder, A.J.B. (1995) Reversibility and mechanism of bacterial adhesion. Colloids and Surfaces B: Biointerfaces 4(1), 5-22.

Rillig, M.C., Wagner, M., Salem, M., Antunes, P.M., George, C., Ramke, H.-G., Titirici, M.-M. and Antonietti, M. (2010) Material derived from hydrothermal carbonization: Effects on plant growth and arbuscular mycorrhiza. Applied Soil Ecology 45(3), 238-242.

Román, S., Nabais, J.M.V., Laginhas, C., Ledesma, B. and González, J.F. (2012) Hydrothermal carbonization as an effective way of densifying the energy content of biomass. Fuel Processing Technology 103(0), 78-83.

Rouquerol, J., Avnir, D., Fairbridge, C.W., Everett, D.H., Haynes, J.H., Pernicone, N., Ramsay, J.D.F., Sing, K.S.W. and Unger, K.K. (1994) Recommendations for the characterization of porous solids. Pure and Applied Chemistry 66(8), 1739-1758.

Sevilla, M., Fuertes, A.B. and Mokaya, R. (2011) High density hydrogen storage in superactivated carbons from hydrothermally carbonized renewable organic materials. Energy & Environmental Science 4(4), 1400-1410.

Shen, C.Y., Wang, L.P., Li, B.G., Huang, Y.F. and Jin, Y. (2012) Role of Surface Roughness in Chemical Detachment of Colloids Deposited at Primary Energy Minima. Vadose Zone Journal 11(1).

Steinbeiss, S., Gleixner, G. and Antonietti, M. (2009) Effect of biochar amendment on soil carbon balance and soil microbial activity. Soil Biology and Biochemistry 41(6), 1301-1310.

Sun, K., Ro, K., Guo, M., Novak, J., Mashayekhi, H. and Xing, B. (2011) Sorption of bisphenol A, 17α-ethinyl estradiol and phenanthrene on thermally and hydrothermally produced biochars. Bioresource Technology 102(10), 5757-5763.

Tian, G., Gu, Z.J., Liu, X.X., Zhou, L.J., Yin, W.Y., Yan, L., Jin, S., Ren, W.L., Xing, G.M., Li, S.J. and Zhao, Y.L. (2011) Facile Fabrication of Rare-Earth-Doped Gd2O3 Hollow Spheres with Upconversion Luminescence, Magnetic Resonance, and Drug Delivery Properties. Journal of Physical Chemistry C 115(48), 23790-23796.

Titirici, M.-M., White, R.J., Falco, C. and Sevilla, M. (2012) Black perspectives for a green future: hydrothermal carbons for environment protection and energy storage. Energy & Environmental Science 5(5), 6796-6822.

Titirici, M.M., Thomas, A., Yu, S.H., Muller, J.O. and Antonietti, M. (2007) A direct synthesis of mesoporous carbons with bicontinuous pore morphology from crude plant material by hydrothermal carbonization. Chemistry of Materials 19(17), 4205-4212.

Truesdail, S.E., Lukasik, J., Farrah, S.R., Shah, D.O. and Dickinson, R.B. (1998) Analysis of Bacterial Deposition on Metal (Hydr)oxide-Coated Sand Filter Media. Journal of Colloid and Interface Science 203(2), 369-378.

Tusi, M.M., Brandalise, M., Polanco, N.S.d.O., Correa, O.V., da Silva, A.C., Villalba, J.C., Anaissi, F.J., Neto, A.O. and Spinacé, E.V. (2013) Ni/Carbon Hybrid Prepared by Hydrothermal Carbonization and Thermal Treatment as Support for PtRu Nanoparticles for Direct Methanol Fuel Cell. Journal of Materials Science & Technology (0).

Wang, Q., Huang, Y., Miao, J., Wang, Y. and Zhao, Y. (2012) Hydrothermal derived Li2SnO3/C composite as negative electrode materials for lithium-ion batteries. Applied Surface Science 258(18), 6923-6929.

Yazdanbakhsh, M., Lotfian, N. and Tavakkoli, H. (2009) Synthesis and characterization of two novel trinuclear oxo-centered, of chromium and iron complexes containing unsaturated carboxylate bridging ligand. Bulletin of the Chemical Society of Ethiopia 23(3), 463-466.

Zaccheo, P., Cabassi, G., Ricca, G. and Crippa, L. (2002) Decomposition of organic residues in soil: experimental technique and spectroscopic approach. Organic Geochemistry 33(3), 327-345.

Zhang, Z.B., Cao, X.H., Liang, P. and Liu, Y.H. (2013) Adsorption of uranium from aqueous solution using biochar produced by hydrothermal carbonization. Journal of Radioanalytical and Nuclear Chemistry 295(2), 1201-1208.

Chapter 5: Removal of *Escherichia coli* from saturated sand columns with intermittent operation supplemented with hydrochar derived from sewage sludge

This chapter is based on:

Chung, J.W., Edewi, O.C., Foppen, J.W., Gerner, G., Krebs, R. and Lens, P.N.L. (submitted) Removal of *Escherichia coli* from saturated sand columns with intermittent operation supplemented with hydrochar derived from sewage sludge.

Abstract

Hydrothermal carbonization (HTC) technology provides a simple, low cost, energy efficient and environmental friendly means to convert various types of waste biomass such as sewage sludge into a carbon rich by-product referred to as hydrochar, which can serve as low-cost adsorbent. We evaluated the *Escherichia coli* (*E. coli*) removal efficiency of a hydrochar derived from stabilized sewage sludge with 10 cm sand columns supplemented with 1.5% (*w/w*) hydrochar. In addition, we evaluated *E. coli* removal in columns of 50 cm using an intermittent flushing regime for 30 days. The raw hydrochar supplement in a sand column was not effective in water treatment for *E. coli* removal. To increase the adsorptive performance, the hydrochar was pre-treated with potassium hydroxide. The highest *E. coli* removal efficiencies were obtained from activated hydrochar amendments: 90.3% for small columns and 99.7% for large columns. This improvement could be attributed to an increase in hydrophobicity of hydrochar observed after KOH activation. For 30 days of intermittent flushing, the mean removal efficiencies of *E. coli* for the sand, raw and activated hydrochar amended columns were 36.5% ± 10.1 (n=60), 24.4% ± 10.5 (n=56) and 91.2% ± 7.5 (n=60), respectively. The effect of idle time on the *E. coli* removal efficiency was only significant in the sand column ($P < 0.05$) resulting in 55.2% of the total removal efficiency. This research suggests that it is possible to utilize hydrochar derived from sewage sludge in water treatment for the removal of bacterial contaminants.

5.1 Introduction

Hydrothermal carbonization (HTC), a novel thermal treatment technology, has emerged as an effective waste conversion and treatment process. HTC, also known as "wet pyrolysis", involves the heating of organic feed stock at subcritical water temperatures (180 - 350°C) under autogenous pressures to produce water soluble organics and a carbonaceous solid called hydrochar (Libra et al., 2011). The attraction of the HTC process is primarily due to its minimal environmental impact, simplicity, cost effectiveness, low greenhouse gas emission, and energy efficiency (Titirici et al., 2012). The first HTC experiment was carried out in 1913 by Bergius as a means to stimulate the natural coalification of organic matter under laboratory conditions (Funke and Ziegler, 2010). Since then, a lot of research has been done using HTC to convert feedstock ranging from pure substances (cellulose and glucose) to more complex ones like paper or empty fruit shells resulting in promoting the integration of carbon in the solid product hydrochar (Lu et al., 2012). The main advantage of HTC compared to other pyrolysis methods is that it can convert biomass inputs with high moisture content into high yield solid carbon materials with minimal energy requirements for both the process itself and subsequent drying of the solid product. This allows for a rather wide range of continuously generated, renewable residues and waste materials ranging from animal wet manure, municipal sewage sludge and waste activated sludge to human faeces and fish farm waste to be employed as potential feed stocks (Libra et al., 2011).

The presence of pathogenic microorganisms, heavy metals and a wide range of organic pollutants in sewage sludge has limited the use of traditional disposal routes due to the possible adverse health impacts to humans, animals and the environment (Fytili and Zabaniotou, 2008). Recently, HTC has been reported as a cost-effective, and eco-friendly environmental management tool to tackle the disposal challenge of sewage sludge generated from wastewater treatment plants (Escala et al., 2013; He et al., 2013; Parshetti et al., 2013). Also, it is an effective means of converting the high moisture organic rich sewage sludge into carbonaceous products with high adsorption capacity, which could replace commercially produced activated carbon for water purification (Smith et al., 2009). Application of sewage sludge derived hydrochar (product of hydrothermal carbonization) as a potential low-cost adsorbent in water treatment is relatively new. Few research studies have reported their use in the removal of abiotic contaminants like heavy metals and organic substances (Alatalo et al., 2013). However, their use for fecal bacteria removal has not been investigated. The main objective of this research work is, therefore, to evaluate the performance of hydrochar derived from sewage sludge for *E. coli* removal in sand filters and to assess its potential for practical field applications.

5.2 Materials and Methods

5.2.1 *Escherichia coli* suspension

E. coli strain UCFL-94 used in the experiment was originally collected from a grazing field as reported in a previous research work (Lutterodt et al., 2009). This strain was reported to have a relatively low sticking efficiency. UCFL-94 was enriched in 50 mL nutrient broth (OXOID, Basingstoke, Great Britain) at 37 °C for 24 h with agitation at 150 rpm on an orbital shaker. The *E. coli* stock was refreshed weekly. The influent for column experiments was prepared by introducing the *E. coli* stock into artificial groundwater (AGW). The AGW was prepared by dissolving 526 mg / L $CaCl_2.2H_2O$ and 184 mg / L $MgSO_4.7H_2O$ in demineralized (DI) water, and buffered by adding 8.5 mg / L KH_2PO_4, 21.75 mg / L K_2HPO_4 and 17.7 mg / L Na_2HPO_4. The pH and electrical conductivity of the AGW ranged from 6.6-6.8, and from 1012-1030 µS/cm, respectively. The AGW feed water seeded with *E. coli* was prepared and stored at room temperature for at least 24 h to facilitate the adjustment of the bacteria in AGW prior to use. The *E. coli* concentration in AGW did not significantly change within at least 4 days (data not shown). Therefore, the die-off was neglected. The concentration of *E. coli* in the influent was maintained at ~10^6 CFU / mL for small column experiments and ~10^3 CFU / mL for large column experiments. Throughout the experiments, the concentration of *E. coli* was determined by the conventional plate counting method (APHA, 1998) using Chromocult agar plates (ChromoCult® Coliform Agar, Merck, Germany). After the incubation of plates at 37 °C for 24 h, the colonies formed on the plates were enumerated using a Colony Counter (IUL, Barcelona, Spain).

5.2.2 Hydrochar

The hydrochar used in this study was obtained from Zurich University of Applied Sciences (ZHAW, Wädenswil, Switzerland). Briefly, stabilized sewage sludge collected from a wastewater treatment plant was hydrothermally carbonized with supplements of sulphuric and acetic acid. The process time was close to 5 h at a median temperature of 210 °C and a pressure of 21 to 24 bar. The slurry produced was cooled to 20 °C and then mechanically dewatered using a membrane filter press, which produces caked solid hydrochar blocks. The caked hydrochar was ground to fine powder using a mortar and pestle. The hydrochar powders were washed by several rounds of suspending in DI water followed by centrifugation at 2,700 g for 3 min (Hermle 236 HK). The KOH activation of washed raw hydrochar was carried out to enhance the *E. coli* removal efficiency (Chung et al., 2014). Washed hydrochar particles were introduced into 1 M KOH solution at a concentration of 5 g hydrochar (dry weight) / L. The hydrochar-KOH suspension was stirred for 1 h at room temperature, and subsequently washed with DI water as described earlier to achieve a neutral pH (Chung et al., 2014). The optimal concentration of KOH was chosen based on preliminary screening tests (data not shown). The concentration of hydrochar in each suspension was measured by drying a portion of the suspension at 105 °C overnight, and the hydrochar stocks were stored at 4 °C.

5.2.3 Material characterization

Zeta potential

The zeta potential values of *E. coli* and hydrochar were determined as a function of pH (4 - 10 for hydrochar and 5.5 - 8.5 for *E. coli*) using a Zetasizer Nano ZS (Malvern, UK) equipped with a MPT-2 pH auto-titration unit. Prior to the measurements, all test materials were conditioned by repetitive washing in AGW. Test suspensions were diluted until an adequate attenuator (6 - 8) was selected by the instrument.

Elemental composition

Elemental composition of the hydrochar was determined by X-ray fluorescence (XRF) using a SPECTRO-XEPOS (SPECTRO, Kleve, Germany) dispersive spectrometer on pressed powder pellets. The spectrometer was equipped with a 10 mm^2 Si-Drift Detector with Peltier cooling and a spectral resolution (FWHM) at Mn Ka $\leq$ 155 eV for determination in the element range of Na - U (SPECTRO, 2014). Samples were prepared by milling hydrochar into a powder of grain size < 100 μm with a mixer mill MM 400 (Retsch, Haan, Germany) for 5 min and a frequency of 25 s^{-1}. Then, 4 g powder was mixed with 0.9 g Licowax C micro powder PM (Clariant, Muttenz, Switzerland) and pressed into pellets of 32 mm diameter with 15 tons pressure. The prepared pellets were measured in the SPECTRO-XEPOS spectrometer with the TurboQuant-screening method. Each powder pellet was set for one exposure on each side (two repetitions) and the average results were recorded.

Additionally, carbon (C), hydrogen (H), nitrogen (N) and oxygen (O) were determined by dry combustion using a TruSpec analyzer (LECO, St. Joseph, Michigan, USA). For C, H, N and O all samples were first oven-dried at 105 °C until their weight was constant and then milled into a powder with the mixer mill for 5 min and a frequency of 25 s^{-1}. Then, 100 mg sample material was combusted at 950 °C and recorded by a TruSpec CHN Macro Analyzer. Furthermore, 3 mg sample material was combusted at 1300 °C with the additional high-temperature pyrolysis furnace TruSpec Micro Oxygen Module to identify the O content. The analyses were carried out in duplicate and the mean value was recorded.

Surface functional groups

Surface functional groups and their alteration have been examined by Photoacoustic Fourier transform infrared spectroscopy (FTIR-PAS). Infrared spectra (4000 - 400 cm^{-1}) were recorded on a Tensor 37 FTIR spectrometer (Bruker Optics, Fällanden, Switzerland) equipped with a photoacoustic optical cantilever microphone PA301 detector (Gasera, Turku, Finland). An average determination of 32 single spectra was carried out after analyses to improve the signal-to-noise ratio. Additionally, a CO_2 spectrum was recorded for adjustment by subtracting the CO_2 spectrum from the spectra of the samples. Prior to FTIR-PAS analysis, samples had been dried at 105 °C for 1 h and then stored in an exsiccator at room temperature. The PA301 sample cell was purged and refilled with helium gas of 99.999 % purity.

Specific surface area and pore size distribution

Adsorption analyses were performed to calculate the surface area and pore size distribution of the hydrochar variants using an Autosorb-iQ automated gas sorption analyzer (Quantochrome, Boynton Beach, Florida, USA). As a first step, the samples were out-gassed under dynamic vacuum at 120 °C to evaporate remnant water and organics. Hydrochar samples were analyzed using N_2 as the adsorbing gas at 77 K (-196 °C). The surface area was evaluated using multi-point Brunauer–Emmett–Teller (BET) analysis. Pore size distribution in the range of micro- and mesopores was determined by density functional theory (DFT) using the quench solid DFT (QSDFT) method (Neimark et al., 2009). Macropores could not be analyzed with this method.

Surface morphology

The surface morphology of raw and activated hydrochar was determined by Scanning Electron Microscopy (SEM) using a Quanta 250 FEG (FEI, Hillsboro, Oregon, USA). The hydrochar sample was placed onto an adhesive carbon tape on an aluminium stub followed by flushing away detached particles with nitrogen gas of 99.999% purity.

Hydrophobicity

The hydrophobicity of two types of hydrochar was measured by static contact angle measurements using a DSA100 (KRÜSS, Hamburg, Germany). Flat surfaces of hydrochar samples were prepared in identical procedures used for XRF analysis (see 2.3.2) except for an addition of Licowax C micro powder (Jeong et al., 2009). The contact angle analyses were performed in triplicate.

5.2.4 Column experiments

Experimental set-up

Quartz sand of 99.1% purity was used as a supporting matrix for column experiments (Kristall quartz-sand, Dorsilit, Germany). A D50 value of 0.79 mm and a U value (D10 / D60) of 1.81 were obtained from a cumulative grain size mass distribution curve determined from sieve analysis of the sand (Matthess et al., 1991). The sand was washed with 5% HCl to remove impurities, rinsed with demineralized water till the washing water pH became close to DI water and oven-dried at 105°C. In this research, two types (small and large) of columns were used: small columns of 10 cm Omnifit borosilicate glass (250 mm x 25 mm inner diameter; Diba industries, Cambridge, United Kingdom) and large columns of 50 cm acrylic poly vinyl chloride (PVC) pipes with an inner diameter of 5.6 cm. Both types of columns were assembled with appropriate tube fittings and connectors. The hydrochar particles were mixed thoroughly with sand to arrive at a 1.5% (*w/w*) hydrochar concentration. When the hydrochar-sand mixture was loaded into the columns, manual tapping and agitation of the columns was done to ensure proper packing and to prevent channelling. In order to minimize air entrapment in the column, packing materials were carefully compacted using a glass bar or a steel rod during the loading process. The columns were connected to a Masterflex pump (model 77201-60, USA) and washed overnight with DI water to remove residual impurities. An upward flow rate of 1 mL (0.2 cm) / min (small columns) or 33.3 mL (1.35 cm) / min (large columns) was employed in the column experiments.

The columns were kept in a vertical position using tripod stands for the small columns, and a manually constructed steel frame rack for the large columns. In order to determine the pore volume (PV) and ensure the stability of aqueous flow in the columns, chloride tracer tests were carried out prior to *E. coli* flushing. The tracer test comprised of a loading phase with the tracer solution (0.02 M NaCl) and deloading phase with DI water. The chloride concentration in the effluent was determined using ion chromatography (ICS-1000, Dionex, USA).

Small column experiments

The *E. coli* removal efficiency in the small columns was determined in duplicate experiments. Each experiment consisted of flushing with 50 mL AGW seeded with *E. coli* followed by flushing with 50 ml of *E. coli* free AGW, respectively. An upward flow rate of 1 mL / min (0.2 cm / min) and sampling frequency of 5 min were applied. The vertical hydrochar weight distribution in the column after each *E. coli* breakthrough test was determined by excavating the 10 cm column from ten sections at 1 cm per section. Excavated packing materials were dried overnight at 105 °C and incinerated subsequently at 500 - 520 °C for 30 min. The amount of residual hydrochar and sand-hydrochar ratio were estimated from the weight loss difference after incineration. Freshly prepared identical hydrochar samples were treated in the same way to serve as a standard to calculate the incinerated proportion of hydrochar.

Large column experiments

The large column experiments were carried out using 3 pairs of columns; each pair was packed either with sand, raw or activated hydrochar. The intermittent daily flushing of *E. coli* seeded artificial groundwater was performed for 30 days. On a weekly basis, the flushing regime consisted of 5 consecutive daily flushing followed by 2 days of pause so that 24 h of idle time was given for 4 measurements, and 72 h for 1 measurement per week. On each day of flushing, 1 L of the influent (~2 PV, calculated from the chloride tracer test, data not shown) was fed into each column, and the effluent was collected in 10 samples with a sampling volume of 100 mL.

When the flushing started, the first PV of influent expelled the pore water out of column media and the second PV of influent was stored in the column bed until the next flushing. The first 5 effluent samples (1 PV) were the pore water stored in previous flushing, and were regarded to reflect the effect of idle time. The successive 5 effluent samples indicated the direct removal of *E. coli* when passing the column bed. The statistical differences of average removal efficiencies from the 3 types of large column set-ups were identified by One-way ANOVA and Tukey's honestly significant difference test ($p < 0.05$). All statistical analyses were performed by IBM SPSS Statistics for Windows Version 22.0 (Armonk, New York, USA).

5.3 Results

5.3.1 Material characterization

Zeta potential

The zeta potential of *E. coli* strain UCFL-94, raw and activated hydrochar samples applied for the column experiments were negative in the pH range of 6.6 - 6.8 (Figure 5.1). KOH activation of the raw hydrochar slightly increased the zeta potential values. While the zeta potential values for the raw hydrochar in the 6.6-6.8 pH range were around -15 mV, for activated hydrochar the zeta potential was -13 mV. Also, *E. coli* UCFL-94 had a negative zeta potential of around -11 mV.

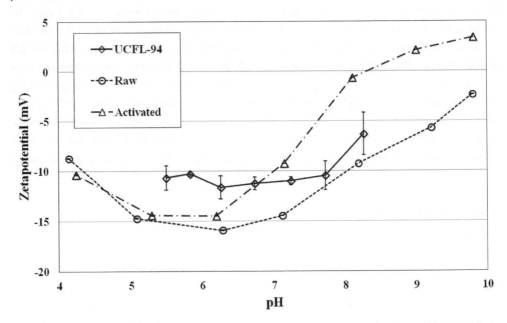

Figure 5.1: Zeta potential of the hydrochar and *E. coli* strain over a function of pH. The lines represent the mean zeta-potential value and error bars indicate the standard deviation.

Elemental composition

The elemental composition of the raw and activated hydrochar were obtained using XRF and a TruSpec analyzer (Table 5.1). The activation with KOH induced a 0.4 % increase of potassium (K) which confirms the sorption of K to the hydrochar surface. The increases of calcium (Ca), magnesium (Mg) and iron (Fe) were most probably the result of impurities contained in KOH. The decreases of aluminium (Al) and phosphorus (P) were due to the washing out in the activation process. The changes in C, H, N and O constitution were not considered as significant.

Table 5.1: Elemental composition (%) of hydrochar variants

	[a]C	H	N	O	[b]Ca	Mg	Al	K	Fe	P
Raw	28.6	3.6	2.0	22.3	5.1	0.9	2.6	0.5	5.0	4.5
Activated	29.4	3.7	1.9	21.1	5.8	1.0	2.0	0.9	5.5	2.6

[a] Obtained from TruSpec analyzer (C, H, N and O) or [b] XRF (Ca, Mg, Al, K, Fe and P)

Surface functional groups

The KOH activation did not significantly alter the qualitative composition of surface functional groups. Similar peak developments were observed from raw and activated hydrochar (Figure 5.2). The spectral fine structure of the FTIR-PAS measurement in the range between 1700 and 1300 cm^{-1} and the bands at 3800 cm^{-1} refer to an artefact of residual water in the samples. The bands at around 2930 - 2850 cm^{-1} correspond to aliphatic C-H stretching vibrations (Sevilla and Fuertes, 2009; Baccile et al., 2013; Parshetti et al., 2013). Bands at 1600 cm^{-1} can be attributed to aromatic C=C stretching vibrations (Sevilla and Fuertes, 2009; Kumar et al., 2011). Aliphatic CH_2 scissoring vibrations refer to the band at 1446 cm^{-1} (Bansal and Goyal, 2005; Silverstein et al., 2005). Activated samples show a slight vibration shift at the bands of 1446 - 1400 cm^{-1}, derived from OH deprotonation at the sample surface. A visible difference occurred in the bands in the 1110-1010 cm^{-1} region, where the samples treated with KOH showed less intensity at C-O stretching vibrations (Bansal and Goyal, 2005; Kumar et al., 2011; Parshetti et al., 2013). Out-of-plane bending vibrations of aromatic C-H bonds can be assigned to the bands at 780 cm^{-1} (Sevilla and Fuertes, 2009; Baccile et al., 2013).

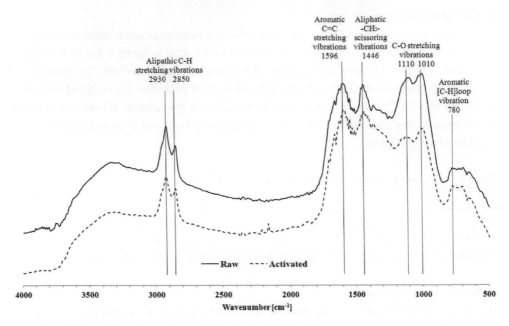

Figure 5.2: FTIR-PAS spectra of raw and activated hydrochar

Specific surface area and pore size distribution

The activation with KOH decreased the BET surface area of hydrochar from 25.3 to 18.5 m^2/ g. The QSDFT pore size (differential pore volume) distribution results are shown in Figure 5.3, which illustrates a constitution of pore volume ascribed to specific pore widths. A negligible pore volume was observed in the micro- and mesopores range. Above a pore width of 15 nm, differences between raw and activated hydrochar were observable. This upper section represents a decrease in surface area after KOH treatment.

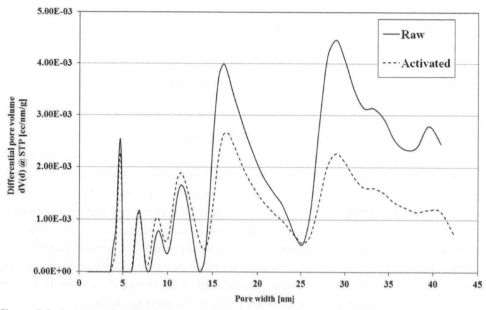

Figure 5.3: Pore size distribution curves obtained from QSDFT analysis

Surface morphology

In order to understand the physical alteration of KOH activation on the surface of raw hydrochar, SEM analyses were carried out. KOH activation induced no significant changes of surface morphology (Figure 5.4). SEM images from both raw and activated hydrochar showed relatively rough surfaces comprising of a macroporous structure (pore width > 50 nm) which might be effective for the attachment of *E. coli* cells having a size of ~2 µm (measured by Zeta-sizer nano, Malvern, data not shown).

Figure 5.4: SEM images of raw (left) and activated (right) hydrochar

Hydrophobicity

The contact angle analyses on raw and activated hydrochar indicated that both materials had hydrophobic surfaces. KOH activation increased the hydrophobicity of hydrochar. The contact angle of raw hydrochar was 126.5 ± 2.9 ° (average of triplicate ± standard deviation), while the contact angle of activated hydrochar was 135.4 ± 4.7 °.

5.3.2   *E. coli* flushing test

The results of breakthrough analysis in small column experiments

The BTCs from small column experiments are presented in Figure 5.5. A clear trend was observed in all BTCs which started with a rising limb, followed by a plateau phase, and finally disappeared with a declining limb. The addition of raw hydrochar in the sand column induced early breakthrough of *E. coli*. While the rising limb of the raw hydrochar amended column started at 10 min, the one from the sand column started at 15 min. Also the declining limb of the raw hydrochar amended column appeared 5 min earlier than the sand column. The transport of *E. coli* could be facilitated by a decrease of pore space in the sand media by filling with the raw hydrochar amendment. Both the sand and raw hydrochar amended column showed similar C/C_0 ratios (~0.9) in the plateau phase, which illustrated the insignificant effect of raw hydrochar in *E. coli* removal. In contrast, the maximum C/C_0 ratio in case of activated hydrochar supplement was only 0.1. The mean removal efficiencies were 9.2, 9.6 and 90.1% for sand, raw hydrochar amended and activated hydrochar amended columns, respectively.

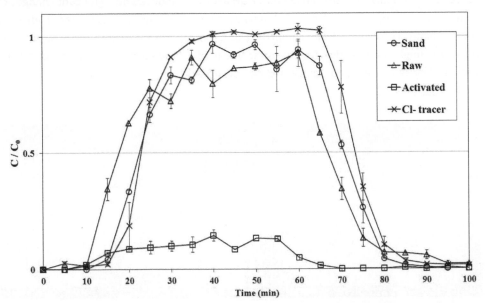

Figure 5.5: Breakthrough curves of *E. coli* from small column experiments carried out at a flow rate of 1 mL (0.20 cm) / min. The solid lines represent the mean C/C_0 value and error bars indicate the individual data points.

5.3.3   *E. coli* removal efficiency in large column experiments

E. coli removal performances for duplicate columns of sand, raw and activated hydrochar amendment are illustrated in Figure 5.6. The columns with activated hydrochar amendment showed the highest removal efficiency of 99.7% on the first day, and the removal performance gently declined during the experimental period, finally ending up at 78.9% on day 30. In contrast, the *E. coli* removal performance of sand and raw hydrochar amended columns were similar and fluctuated between 11.4 and 57.2 %.

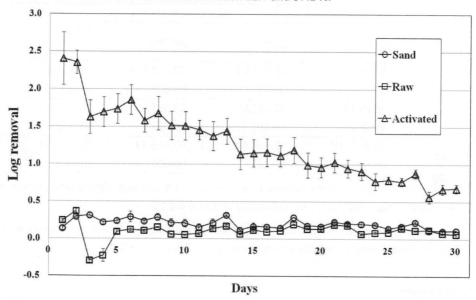

Figure 5.6: *E. coli* removal efficiencies from large columns during 30 days of flushing with a flow rate of 33.3 mL (1.35 cm) / min. The solid lines represent the mean values of duplicate columns and error bars indicate the individual data points

The overall *E. coli* removal performance and the effect of idle time in the large column experiments are summarized in Table 5.2. In all conditions investigated, the columns with activated hydrochar showed superior performance to sand columns amended with raw hydrochar. While the activated hydrochar columns showed 91.2 % of the average total removal efficiency during 30 days of operation, it was only 24.4 % and 36.5 % for raw hydrochar and sand columns, respectively. The idle time given to sand columns improved the *E. coli* removal performance. While the direct removal efficiency of *E. coli* (retention of *E. coli* passing through the column bed, represented by the first PV of effluent) was 17.2 %, the removal efficiencies for those experiments including idle time were 52.1 % for 24 hr and 66.9 % for 72 hr. In contrast, the effect of 24 hr idle in the columns with either raw or activated hydrochar supplements was negligible. The extended idle time of 72 hr given to raw hydrochar-amended columns increased the *E. coli* removal by 14.8 %. Throughout the experiments, the *E. coli* removal efficiency of activated hydrochar-amended columns was independent from the idle time. Excluding the effect of idle time, the direct removal performance of *E. coli* observed in the second PV effluent was similar in both sand and raw hydrochar amended columns. This corresponded well with the results from small column experiments.

Table 5.2: Removal efficiency (average % ± standard deviation) of *E. coli* in large column experiments during 30 days of intermittent flushing in duplicate columns of each treatment.

Content	Pore volume		Second	Total
	First			
	24 hr idle	72 hr idle		
Sand	52.1±13.3	66.9±14.7	17.2±8.64§[a]	36.5±10.1
	(n=46)	(n=12)	(n=60)	(n=60)
§Raw	23.0±17.3†[b]	35.1±12.9	20.0±12.5 §†	24.4±10.5†
	(n=44)	(n=10)	(n=56)	(n=56)
Activated	92.8±5.0‡[b]	92.6±8.0 ‡	90.0±9.1‡	91.2±7.5‡
	(n=46)	(n=12)	(n=60)	(n=60)

[a] within each column values followed by the symbol § are not significantly different using Tukey's honestly significant difference test at p < 0.05.

[b] within each row values followed by the same symbol † or ‡ are not significantly different using Tukey's honestly significant difference test at p < 0.05.

5.4   Discussion

5.4.1   Effect of KOH activation of hydrochar on *E. coli* removal

The adsorption performance of an adsorbent is primarily based on its surface area that provides many adsorptive sites and internal pore structure (Unur, 2013). The attachment of *E. coli* on surfaces can be determined by several interactions including repulsive and attractive forces (Foppen and Schijven, 2006). Desirable characteristics for an *E. coli* adsorbent would be a positive (less negative) surface charge, highly porous and hydrophobic material. Previous researches investigated the use of plant material-derived hydrochar for heavy metal or *E. coli* removal from aqueous solution, where KOH treatment at room temperature enhanced the surface roughness resulting in improved contaminant removal (Regmi et al., 2012; Chung et al., 2014). Surprisingly, the BET and SEM assessments in this research could not explain the improvement in *E. coli* removal performance induced by KOH activation. No significant modification was observed in surface morphology, and, moreover, the specific surface area of hydrochar due to KOH treatment decreased by 27%. We speculate that the decrease in specific surface area may have been due to the transformation of the micro- and mesopores on the raw hydrochar surface to macropores during the KOH activation and subsequent water-washing process. Regarding the size of *E. coli* (~2 μm), BET analysis for micro (< 2 nm) and mesoporous (2-50 nm) surface area may not be a direct explanation for the *E. coli* removal performance. Further quantitative assessment of the macroporous surface area is recommended in order to have a better understanding of the *E. coli* removal mechanism.

Since the assessments of the elemental composition, surface charge and surface functional groups demonstrated only minor alterations resulting from the KOH activation, a possible explanation for the increased removal of *E. coli* in our research can be found from the increased hydrophobicity resulting in improved hydrophobic attraction between *E. coli* and the hydrochar surfaces. Hydrochar is characterized by a hydrophobic core and a stabilizing hydrophilic outer surface (Jarrah et al., 2003; Sevilla and Fuertes, 2009; Roldán et al., 2012). The activation of raw hydrochar with a strong alkaline solution would have induced more hydrophobic sites by removing the hydrophilic surface layer formed in the repolymerization / recondensation of water soluble components in HTC process (Kumar et al., 2011).

5.4.2 Effect of idling time on *E. coli* removal efficiency

The effect of the idle time, which refers to the pause period between daily flushing of the columns with the pores saturated, has a significant impact on the bacterial removal efficiency in sand filtration units operated on an intermittent basis (Elliott et al., 2008). Our observation from the large sand columns (see Table 5.2) agrees with this report that bacterial surrogates stored in the filter media were attenuated during the idle time. The mechanisms for the bacterial inactivation during the idle time of intermittently operated sand filters are still ambiguous. This attenuation could be attributed to facilitated attachment on sand surfaces. We speculate that *E. coli* in the pore water had more chances to attach on the sand surfaces during the idle time due to Brownian diffusion (Li et al., 2008) and motile mobility (Maeda et al., 1976). In contrast, the effect of the idle time was negligible in either raw or activated hydrochar columns. The presence of carbonaceous surfaces in the sand bed can have provided better survival conditions to the *E. coli* cells such as protection from external stresses and nutrient sources (Gerba and Schaiberger, 1975; Burton et al., 1987; Sherer et al., 1992; Howell et al., 1996). More investigations on the interaction between *E. coli* and sand / hydrochar surfaces under static conditions are necessary in order to generate a better understanding of the effect of idle time on bacterial inactivation.

5.4.3 Potential of application in HTC-sand filter for pathogen removal

Provision of safe drinking water can be provided either through centralized piped water treatment and/or through point-of-use (POU) water treatment technologies. POU household water treatment technologies such as the biosand filters (BSF) have been categorized as low-cost and an effective means for providing safe water to poor rural communities (Sobsey et al., 2008). However, the insufficient performance of BSF during the filter ripening period was pointed out as a limitation for stable supply of biologically safe water. Previous studies on BSF employing comparable experimental conditions to this research reported insufficient *E. coli* removal efficiencies (60 - 70%) in the first 3 weeks (Stauber et al., 2006; Elliott et al., 2008). Since the activated hydrochar amended columns showed superior removal efficiencies (96.5%) in the same period, supplements of activated hydrochar into BSF could be a viable option.

The effect of hydrochar supplements on sand filter maturation remains unexplored. Filter maturation and biofilm formation have been reported as of paramount importance in the performance of slow sand filters and other modified systems like the BSF. Though they were operated for 30 days, which is close to the 4-6 weeks reported for filter maturation in a BSF (Ahammed and Davra, 2011), the *E. coli* removal efficiency in our large column experiments trend does not suggest filter maturation and biofilm formation. Biofilm formation and filter maturation may be stimulated by the use of typical surface water quality as in river water or canal water with heterogeneous microbial populations instead of the AGW used in this study. More research is needed to assess the effect of biofilm formation and filter maturation on the removal efficiency of pathogens in hydrochar supplemented sand columns.

5.5 Conclusions

This study demonstrated the potential use of hydrochar derived from sewage sludge in water treatment as a capable low-cost pathogen barrier. Cold alkali activation of hydrochar increased hydrophobicity of hydrochar by removing hydrophilic substances from the surface of hydrochar resulting in increased *E. coli* removal. While the raw hydrochar supplement (1.5 %, *w/w*) was not effective (24.4 %), activated hydrochar supplement in the sand column with 50 cm bed height showed improved *E. coli* removal efficiency of 91.2 % during 30 days of intermittent operation. The mitigation of *E. coli* in the filter idle time was only obvious in the sand columns resulting in the total removal efficiency of 55.2%. Pathogen removal in water treatment is a new approach in hydrochar utilization. Extensive following up investigations are required for practical implementation. Our research could encourage future researches on related topics including biofilm development in hydrochar-supplemented sand filtration units and fate of adhered microorganisms on hydrochar surface.

Acknowledgements

This research was funded by the Korean Church of Brussels, Mangu Jeja Church (Seoul, Korea), and the Netherlands Ministry of Development Cooperation (DGIS) through the UNESCO-IHE Partnership Research Fund. It was carried out in the framework of the research project 'Addressing the Sanitation Crisis in Unsewered Slum Areas of African Mega-cities' (SCUSA).

5.6 References

Ahammed, M.M. and Davra, K. (2011) Performance evaluation of biosand filter modified with iron oxide-coated sand for household treatment of drinking water. Desalination 276(1), 287-293.

Alatalo, S.M., Repo, E., Makila, E., Salonen, J., Vakkilainen, E. and Sillanpaa, M. (2013) Adsorption behavior of hydrothermally treated municipal sludge & pulp and paper industry sludge. Bioresource Technology 147, 71-76.

APHA (ed) (1998) Standard Methods for the Examination of Water and Wastewater (20th ed.), American Public Health Association, Washington, D.C.

Baccile, N., Weber, J., Falco, C. and Titirici, M.-M. (2013) Sustainable Carbon Materials from Hydrothermal Processes, pp. 151-211, John Wiley & Sons, Ltd.

Bansal, R.C. and Goyal, M. (2005) Activated carbon adsorption (p. 33), CRC Press, Taylor & Francis Group, Boca Raton, FL 33487-2742.

Burton, G.A., Jr., Gunnison, D. and Lanza, G.R. (1987) Survival of pathogenic bacteria in various freshwater sediments. Applied and Environmental Microbiology 53(4), 633-638.

Chung, J.W., Foppen, J.W., Izquierdo, M. and Lens, P.N.L. (2014) Removal of Escherichia coli from saturated sand columns supplemented with hydrochar produced from maize. Journal of Environmental Quality 43(6), 2096-2103.

Elliott, M., Stauber, C., Koksal, F., DiGiano, F. and Sobsey, M. (2008) Reductions of *E. coli*, echovirus type 12 and bacteriophages in an intermittently operated household-scale slow sand filter. Water research 42(10), 2662-2670.

Escala, M., Zumbuhl, T., Koller, C., Junge, R. and Krebs, R. (2013) Hydrothermal Carbonization as an Energy-Efficient Alternative to Established Drying Technologies for Sewage Sludge: A Feasibility Study on a Laboratory Scale. Energy & Fuels 27(1), 454-460.

Foppen, J.W.A. and Schijven, J.F. (2006) Evaluation of data from the literature on the transport and survival of Escherichia coli and thermotolerant coliforms in aquifers under saturated conditions. Water Research 40(3), 401-426.

Funke, A. and Ziegler, F. (2010) Hydrothermal carbonization of biomass: A summary and discussion of chemical mechanisms for process engineering. Biofuels Bioproducts & Biorefining-Biofpr 4(2), 160-177.

Fytili, D. and Zabaniotou, A. (2008) Utilization of sewage sludge in EU application of old and new methods—A review. Renewable and Sustainable Energy Reviews 12(1), 116-140.

Gerba, C.P. and Schaiberger, G.E. (1975) Effect of particulates on virus survival in seawater. Journal - Water Pollution Control Federation 47(1), 93-103.

He, C., Giannis, A. and Wang, J.Y. (2013) Conversion of sewage sludge to clean solid fuel using hydrothermal carbonization: Hydrochar fuel characteristics and combustion behavior. Applied Energy 111, 257-266.

Howell, J.M., Coyne, M.S. and Cornelius, P.L. (1996) Effect of sediment particle size and temperature on fecal bacteria mortality rates and the fecal coliform/fecal streptococci ratio. Journal of Environmental Quality 25(6), 1216-1220.

Jarrah, N., van Ommen, J.G. and Lefferts, L. (2003) Development of monolith with a carbon-nanofiber-washcoat as a structured catalyst support in liquid phase. Catalysis Today 79(1-4), 29-33.

Jeong, S.-B., Yang, Y.-C., Chae, Y.-B. and Kim, B.-G. (2009) Characteristics of the Treated Ground Calcium Carbonate Powder with Stearic Acid Using the Dry Process Coating System. Materials Transactions 50(2), 409-414.

Kumar, S., Loganathan, V.A., Gupta, R.B. and Barnett, M.O. (2011) An Assessment of U(VI) removal from groundwater using biochar produced from hydrothermal carbonization. Journal of Environmental Management 92(10), 2504-2512.

Li, G., Tam, L.K. and Tang, J.X. (2008) Amplified effect of Brownian motion in bacterial near-surface swimming. Proceedings of the National Academy of Sciences of the United States of America 105(47), 18355-18359.

Libra, J.A., Ro, K.S., Kammann, C., Funke, A., Berge, N.D., Neubauer, Y., Titirici, M.M., Fühner, C., Bens, O., Kern, J. and Emmerich, K.H. (2011) Hydrothermal carbonization of biomass residuals: A comparative review of the chemistry, processes and applications of wet and dry pyrolysis. Biofuels 2(1), 71-106.

Lu, X., Jordan, B. and Berge, N.D. (2012) Thermal conversion of municipal solid waste via hydrothermal carbonization: Comparison of carbonization products to products from current waste management techniques. Waste Management 32(7), 1353-1365.

Lutterodt, G., Basnet, M., Foppen, J.W.A. and Uhlenbrook, S. (2009) Determining minimum sticking efficiencies of six environmental Escherichia coli isolates. Journal of Contaminant Hydrology 110(3-4), 110-117.

Maeda, K., Imae, Y., Shioi, J.I. and Oosawa, F. (1976) Effect of temperature on motility and chemotaxis of Escherichia coli. Journal of bacteriology 127(3), 1039-1046.

Matthess, G., Bedbur, E., Gundermann, K.O., Loof, M. and Peters, D. (1991) Investigation on filtration mechanisms of bacteria and organic particles in porous-media .1. Background and methods. Zentralblatt Fur Hygiene Und Umweltmedizin 191(1), 53-97.

Neimark, A.V., Lin, Y., Ravikovitch, P.I. and Thommes, M. (2009) Quenched solid density functional theory and pore size analysis of micro-mesoporous carbons. Carbon 47(7), 1617-1628.

Parshetti, G.K., Liu, Z.G., Jain, A., Srinivasan, M.P. and Balasubramanian, R. (2013) Hydrothermal carbonization of sewage sludge for energy production with coal. Fuel 111, 201-210.

Regmi, P., Moscoso, J.L.G., Kumar, S., Cao, X.Y., Mao, J.D. and Schafran, G. (2012) Removal of copper and cadmium from aqueous solution using switchgrass biochar produced via hydrothermal carbonization process. Journal of Environmental Management 109, 61-69.

Roldán, L., Santos, I., Armenise, S., Fraile, J.M. and García-Bordejé, E. (2012) The formation of a hydrothermal carbon coating on graphite microfiber felts for using as structured acid catalyst. Carbon 50(3), 1363-1372.

Sevilla, M. and Fuertes, A.B. (2009) Chemical and Structural Properties of Carbonaceous Products Obtained by Hydrothermal Carbonization of Saccharides. Chemistry-a European Journal 15(16), 4195-4203.

Sherer, B.M., Miner, J.R., Moore, J.A. and Buckhouse, J.C. (1992) Indicator bacterial survival in stream sediments. Journal of Environmental Quality 21(4), 591-595.

Silverstein, R.M., Webster, F.X. and Kiemle, D.J. (2005) Spectrometric identification of organic compounds, 7th edition (p.82-88), John Wiley & Sons, Hoboken, NJ 07030-577.

Smith, K.M., Fowler, G.D., Pullket, S. and Graham, N.J.D. (2009) Sewage sludge-based adsorbents: A review of their production, properties and use in water treatment applications. Water Research 43(10), 2569-2594.

Sobsey, M.D., Stauber, C.E., Casanova, L.M., Brown, J.M. and Elliott, M.A. (2008) Point of use household drinking water filtration: a practical, effective solution for providing sustained access to safe drinking water in the developing world. Environmental science & technology 42(12), 4261-4267.

Stauber, C.E., Elliott, M.A., Koksal, F., Ortiz, G.M., DiGiano, F.A. and Sobsey, M.D. (2006) Characterisation of the biosand filter for E. coli reductions from household drinking water under controlled laboratory and field use conditions. Water Science and Technology 54(3), 1-7.

Titirici, M.-M., White, R.J., Falco, C. and Sevilla, M. (2012) Black perspectives for a green future: hydrothermal carbons for environment protection and energy storage. Energy & Environmental Science 5(5), 6796-6822.

Unur, E. (2013) Functional nanoporous carbons from hydrothermally treated biomass for environmental purification. Microporous and Mesoporous Materials 168(0), 92-101.

Chapter 6: Removal of rotavirus and adenovirus from artificial ground water using hydrochar derived from sewage sludge

This chapter is based on:

Chung, J.W., Foppen, J.W., Gerner, G., Krebs, R. and Lens, P.N.L. (2015) Removal of rotavirus and adenovirus from artificial ground water using hydrochar derived from sewage sludge. Journal of Applied Microbiology 119(3), 876-884.

Abstract

Aim: To determine the pathogenic virus removal performance of an adsorbent produced from hydrothermal carbonization of sewage sludge.

Methods and Results: The removal of human pathogenic rotavirus and adenovirus was investigated with columns of 10 cm saturated sand with and without amendments of 1.5% (*w/w*) hydrochar. Virus concentrations were determined with reverse transcription (RT) quantitative polymerase chain reaction (qPCR). The experiments with sand showed 1 log removal, while the columns with 1.5 % (*w/w*) hydrochar amendment showed 2 to > 3 log removal for both viruses. Deionized (DI) water flushing into the virus-retaining columns revealed that the secondary energy minimum played a larger role in the attachment of rotavirus onto hydrochar surfaces than adenovirus. Improved virus removal may be attributed to the introduction of hydrophobic and/or meso-macro surface structures of the hydrochar providing favorable attachment sites for viruses.

Conclusions: Hydrochar amended sand beds showed improved virus removal efficiencies exceeding 99.6 % corresponding to 2.4 log removal. The addition of humic acid in the influent did not hinder the adsorptive removal of viruses.

Significance and Impact of the Study: This study suggests that hydrochar derived from sewage sludge can be used as an adsorbent for virus removal in water treatment.

6.1 Introduction

Access to sufficient clean drinking water and adequate sanitation is an essential requirement for the well-being of humans. Currently, 0.9 billion people are living without improved drinking-water sources (WHO and UNICEF, 2010). In spite of the recent decrease of mortality, diarrhoea still remains one of the main causes of child death. Complications with other diseases such as malaria and pneumonia stimulates diarrheal death of children (Kosek et al., 2003). Though the world population connected to improved drinking water and sanitation has increased, the number of people facing risks in less developed regions also increased because of the rapid population growth (Aertgeerts, 2009). Both in developed and developing countries, rotavirus and adenovirus are strongly associated with water borne diarrheal diseases (Gomara et al., 2008; Maunula et al., 2009; Parashar et al., 2009). Due to the strong persistency of rotavirus and adenovirus against external stresses including water/wastewater treatment measures, the development of affordable virus removal techniques is necessary (Crabtree et al., 1997; Pina et al., 1998; Chapron et al., 2000; Carter, 2005; Estes et al., 2008; Xue et al., 2013).

Recently, hydrothermal carbonization (HTC) has been highlighted as a cost-effective tool for producing valuable carbonaceous materials from biowaste (Elaigwu et al., 2014; Kim et al., 2014; Poerschmann et al., 2014). In the HTC process, valuable carbonaceous products can be produced by heating a suspension of feed stock and water to temperatures of 180 – 200 °C at saturated pressure for several hours (Funke and Ziegler, 2010). In contrast to ordinary dry pyrolysis techniques, HTC can convert wet-biomass into carbonaceous materials without an energy-intensive drying process. Accordingly, the applicability of wet biomass initiated the research on various non-traditional potential feed stocks, including wet animal/human excrements, sewage sludge, municipal solid waste, aquaculture and algal residues (Libra et al., 2011). Previous research suggested the potential of hydrochar, the solid product of HTC, in water treatment (Sun et al., 2011; Regmi et al., 2012). Various abiotic contaminants such as heavy metals, radioactive agents, organic dye, endocrine disrupters and phosphorus were efficiently removed from aqueous solutions in water treatment set-ups (Kumar et al., 2011; Sun et al., 2011; Dai et al., 2014; Minani et al., 2014; Parshetti et al., 2014). Chung et al. (2014) further reported an efficient removal of *Escherchia coli* in a sand column set-up supplemented with hydrochar derived from agricultural residue of maize. However, still no one has reported the adsorptive removal of pathogenic viruses using an adsorbent produced by HTC. Therefore, the main objective of this research is to provide an evaluation on the hydrochar derived from sewage sludge as an adsorbent in water treatment for virus removal. As the most significant viruses responsible for diarrheal diseases, the removal of the human pathogenic rotavirus and adenovirus were investigated by carrying out breakthrough analyses of their adsorption onto the hydrochar.

6.2 Methods and materials

6.2.1 Hydrochar and characteristics

The same hydrochar stock as used in experiments for *Escherichia coli* (*E. coli*) retention in soil (Chung et al., submitted) was used in this work. Briefly, stabilized sewage sludge collected from a wastewater treatment plant was hydrothermally carbonized for 5 h at a median temperature of 210 °C and a pressure between 21 to 24 bar. The outcome of the reaction had a form of slurry consisting of hydrochar particles and process water. It was cooled to 20 °C and dewatered through a membrane filter press. The raw hydrochar stock was ground into fine powder using a mortar. The hydrochar powders were washed several times by suspending in DI (deionized) water followed by centrifugation at 2,700 g for 3 min. This washed hydrochar stock was stored at 4 °C until further use.

Multi point Brunauer–Emmett–Teller (BET) analysis showed that the hydrochar had a specific surface area of 25.3 m^2 g^{-1}, which is largely attributed by the formation of mesopores determined by density functional theory (DFT) using the quench solid DFT (QSDFT) method. The zeta-potential value of hydrochar particles dispersed in artificial groundwater (AGW) was - 15 mV at pH 6.8. The static contact angle of a pressed hydrochar disc was 126 °, which was considered to be strongly hydrophobic (Chung et al., submitted).

6.2.2 Virus stock and influent preparation

Stocks of *Rotavirus WA* (RV) with a concentration of $10^{5.8}$ tissue culture infectious dose 50 (TCID$_{50}$) ml^{-1} and *Human adenovirus C* type 2 (HAdV) with a concentration of 10^6 plaque - forming unit (PFU) ml^{-1} were kindly provided by The Netherlands National Institute for Public Health and Environment (RIVM, Bilthoven, The Netherlands). The frozen stocks were thawed in a safety chamber satisfying bio-safety level (BSL) 2 and were aliquoted in Eppendorf vials. These vials were then stored in a freezer (-80 °C) until needed.

The virus removal efficiency of hydrochar was evaluated by performing breakthrough analyses. The influent for a virus removal experiment was prepared by dilution of standard virus stocks into Artificial Ground Water (AGW) at a concentration of ~10^3 TCID$_{50}$ (RV) or PFU (HAdV) ml^{-1}. The AGW was prepared by dissolving 526 mg $CaCl_2.2H_2O$, 184 mg $MgSO_4.7H_2O$, 8.5 mg KH_2PO_4, 21.8 mg K_2HPO_4 and 17.7 mg Na_2HPO_4 in 1 l DI water (Lutterodt et al., 2009). The resulting pH and electrical conductivity (EC) of the AGW were ~ 6.8 and 1000 μS cm^{-1}, respectively. Also, a second type of influent was prepared by the addition of Na-humic acid (ACROS Organics, Belgium) in AGW at a concentration of 18 mg l^{-1} in order to investigate the effect of humic acid on the virus removal performance. It is reported that humic acid in aqueous solutions compete with microbial agents for the same adsorption sites (Schijven and Hassanizadeh, 2000). Therefore, the removal of bacteriophage PRD1 was inhibited (Foppen et al., 2006). The virus-containing influent was stabilized at room temperature for 30 min prior to feeding into columns.

6.2.3 Virus enumeration

The virus concentration was examined by performing RT-qPCR for RV and qPCR for HAdV. The nucleic acids of viruses contained in 100 µl sample were extracted as previously described using chaotropic lysis buffers containing high molarity (5.25 mol l^{-1}) guanidine thiocyanate (GuSCN) (Boom et al., 1990; Boom et al., 1999) and stored at -80 °C. Table 6.1 summarizes the design of primers and probes for HAdV (Hernroth et al., 2002) and RV (Pang et al., 2004). These primers and probes were synthesized by Biolegio (Nijmegen, The Netherlands).

Table 6.1: Nucleic acid sequences of primers and probes used in (RT)-qPCR assays.

		Sequence (5'–3')	Reference
Rotavirus	Forward primer	ACCATCTACACATGACCCTC	Pang et al., (2004)
	Reverse primer	GGTCACATAACGCCCC	
	Probe	6FAM-ATGAGCACAATAGTTAAAAGCTAACACTGTCAA-BHQ1	
Adenovirus	Forward primer	CWTACATGCACATCKCSGG	Hernroth et al., (2002)
	Reverse primer	CRCGGGCRAAYTGCACCAG	
	Probe	6FAM-CCGGGCTCAGGTACTCCGAGGCGTCCT-BHQ1	

The concentration of RV was determined by performing a two-step RT-qPCR assay. First, the RNA extracted from samples was reverse transcribed to cDNA through a reverse transcription process using RevertAid revert transcriptase (Thermo Fisher Scientific, Pittsburgh, Pennsylvania, U.S.), and the cDNA was enumerated using qPCR (Chung et al., 2013). For the determination of HAdV , 4 µl of nucleic acid was added to a qPCR mix consisting of 2.5 µl home-made qPCR buffer (150 mmol l^{-1} Tris - HCl pH 8.2 at 25°C, 300 mmol l^{-1} KCl, 50 mmol l^{-1} $(NH_4)_2SO_4$, 25 mmol l^{-1} $MgCl_2$, and 0.2% BSA), 1 µl dNTP mix, 1 µl (0.5 unit) Taq polymerase (both from GenScript, Piscataway, New Jersey, U.S.), 15.4 µl DEPC treated water, 0.4 µl of forward and reverse primer (both at a 200 nmol l^{-1} final concentration), and 0.3 µl probe (150 nmol l^{-1} final concentration). Then, the samples were exposed to a thermal cycling regime consisting of 95 °C for 10 min, and then 40 cycles of denaturation at 95 °C for 15 s followed by annealing/extension at 60 °C for 60 s. The fluorescence signal was measured at the end of each extension step. All PCR experiments were carried out using a BioRad MJMini™ (Bio-Rad, Hercules, California, U.S.). In order to convert the cycle of threshold cycle values (Cts) obtained from (RT)-qPCR assessments into actual virus concentrations, regression curves of RV and HAdV were constructed in triplicate from a 10-fold dilution series of standard virus stocks: 6 samples with concentrations ranging from $10^{0.8}$ - $10^{5.8}$ $TCID_{50}$ ml^{-1} (RV) or 10 - 10^6 PFU ml^{-1}. The virus concentrations of the samples from column experiments were determined using standard equations derived from these calibration curves.

6.2.4 Column preparation

In this research, the virus removal efficiency of hydrochar was evaluated by comparing two types of sand filtration set-ups using either sand or sand supplemented with hydrochar as packing materials. The columns were prepared as described previously (Chung et al., 2014). Briefly, a stock of quartz sand with 99.1 % purity (Kristall quartz-sand, Dorsilit, Germany) was immersed in 5 % HCl overnight to remove impurities, and rinsed with DI water until obtaining pH values of 6 - 7 in the washing water. Finally, the sand stock was dried at 105 °C in an oven. The sieve analysis on the sand stock obtained a particle size distribution of 0.5 % in the size range ≤ 0.425 mm, 32.2 % in the size range 0.425 ~ 0.560 mm, 41.0 % in the size range 0.560 - 1.1 mm, and 26.4 % in the size range 1.1 ~ 1.6 mm. The packing material for column experiments was prepared by adding the hydrochar slurry into oven-dried sand at a concentration of 1.5 % (dry weight hydrochar/sand). The sand-hydrochar stock was thoroughly mixed and packed in a borosilicate glass column with a 2.5 cm inner diameter and a 10 cm bed height (Omnifit, Cambridge, U.K.).

In the packing process, the column was tapped and agitated regularly to minimize channeling and air entrapment. Also, the sand-hydrochar mixture loaded in the column was carefully compacted using a glass bar throughout the packing process. Then, the column was coupled with appropriate tube fittings and connected to a peristaltic pump (MasterFlex model 77201-60, USA) equipped with a manual rotary valve for feed water selection. The column was washed with DI water overnight to remove residual fines and chemicals. Prior to each flushing experiment, AGW was flushed into the column until the EC value in the effluent became close to that of the feed AGW. An upward flow rate of 1 ml min^{-1} (0.2 cm min^{-1}) was applied in all column flushing experiments.

6.2.5 Design of virus removal experiments

Breakthrough curves (BTC) were obtained from 8 triplicate experimental set-ups composed of two types of viruses (RV or HAdV), two types of packing material (sand or sand-hydrochar) and two types of influent (AGW or AGW-humic acid). For each breakthrough analysis, the loading and deloading of viruses was conducted by flushing 50 ml of influent followed by deloading with 50 ml of virus-free influent with an upward flow rate of 1 ml min^{-1}. After the deloading phase, 100 ml (or 50 ml for sand columns with HAdV flushing) of DI water was flushed through the column in order to determine the role of secondary energy minimum in viral retention in the columns. The idea was that the low ionic strength of DI water (4 - 5 μS cm^{-1}) would increase the double layer repulsion between virions and the surface of the packing media, thereby releasing virions deposited in the secondary energy minimum (Foppen et al., 2007; Chung et al., 2014). We anticipated that with this experiment we could learn about the type of attachment of the virus particles to the collector surfaces. The concentration of the RV or HAdV in the effluent was measured at 5 ml intervals by (RT)-qPCR assay as described above. Every breakthrough experiment was performed in triplicate. All operations handling active viruses were performed in a biosafety cabinet with biosafety level (BSL) 2 qualification. Virus-contaminated apparatuses and the column media were autoclaved at 121 °C for 20 min prior to disposal or cleaning-up.

6.3 Results

6.3.1 Calibration curves

Standard curves were prepared using 10-fold dilutions of the RV or HAdV stock (Figure 6.1). The standard curves were clearly log-linear with R^2-values higher than 0.99. The amplification efficiencies determined from the slope of the calibration curves were 97.7 % for RV and 95.4 % for HAdV, which we considered to be good.

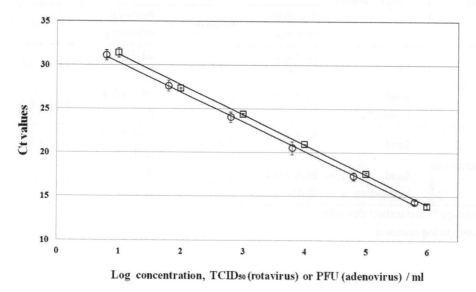

Figure 6.1: Calibration curves for rotavirus and adenovirus. (o) Rotavirus and (□) Adenovirus.

6.3.2 Breakthrough analyses

The BTCs from the experiments performed without humic acid are given in Figure 6.2. The sand columns showed a typical trend of BTC composed of a falling limb, a plateau phase and a rising limb in both RV and HAdV flushing. Commonly, the falling limb began from 10 ml, and arrived to the plateau phase at 20 ml, which was maintained for 40 ml with a little fluctuation around the log removal value (-log C/C_0, LRV) of 1.0 and 1.3. Finally, the rising limb appeared at 60 ml, the virus concentration of the RV arrived to below the detection limit at 75 ml. The rising limb of HAdV also started at 60 ml, however, a minor concentration (LRV 2.2 – 2.8) of HAdV was still observed between 60 and 100 ml while other BTCs showed complete retention of viruses in the column media. This would indicate a slow release of HAdV from the sand media. The total virus removal efficiencies from the sand columns were 92.3 % for RV and 92.7 % for HAdV (Table 6.2). In contrast, the columns with hydrochar supplements showed improved removal of viruses. While the majority of samples appeared to be virus-free, only a few samples showed detectable levels of viruses irregularly. Accordingly, superior virus removal efficiencies were obtained for RV (99.8 %) and HAdV (99.6 %).

Table 6.2: Removal and release of viruses in the breakthrough analysis with deionized water flushing.

Virus	Column media	Type of influent			
		AGW		AGW + humic acid	
		Removal efficiency	Release in DI	Removal efficiency	Release in DI
Rotavirus	Sand	92.3 ± 4.0 [a] (1.1) [b]	3.1 ± 1.1	92.7 ± 2.0 (1.1)	21.1 ± 3.0
	Sand + Hydrochar	99.8 ± 0.3 (2.7)	24.2 ± 4.9	99.8 ± 0.3 (2.7)	16.4 ± 4.5
Adenovirus	Sand	92.7 ± 6.0 (1.1)	5.3 ± 3.9	98.9 ± 0.9 (2.0)	10.3 ± 8.3
	Sand + Hydrochar	99.6 ± 0.4 (2.4)	4.7 ± 2.5	99.9 ± 0.2 (3.0)	3.1 ± 2.9

[a] average % ± standard deviation
[b] average log removal

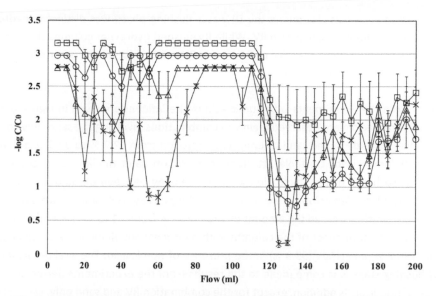

Figure 6.2: Breakthrough curves from artificial ground water flushing: Solid lines represent the mean log removal (-log C/C_0) value and error bars indicate the standard deviation. Note that the deionized water flushing started at 100 ml. The log removal values of samples under the detection limit were presented as C_0. Columns: (×) Rotavirus-Sand, (o) Rotavirus-HTC, (Δ) Adenovirus-Sand and (□) Adenovirus-HTC.

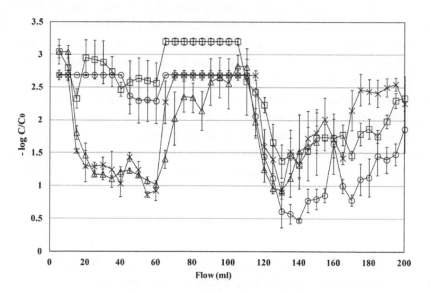

Figure 6.3: Breakthrough curves from humic acid-containing artificial ground water flushing: Solid lines represent the mean log removal (-log C/C_0) value and error bars indicate the standard deviation. Note that the deionized water flushing started at 100 ml. The log removal values of samples under the detection limit were presented as C_0. Columns: (×) Rotavirus-Sand, (o) Rotavirus-HTC, (Δ) Adenovirus-Sand and (□) Adenovirus-HTC.

The release of virus particles during the DI water flushing phase was distinct in hydrochar supplemented columns flushed with RV. While RV-flushed hydrochar columns showed 24.2 % of virus release with the lowest LRV of 0.5, it was 3.1 - 5.3 % for all the other columns with the lowest LRV between 0.9 and 1.3.

The BTCs from the experiments with humic acid in the influent are given in Figure 6.3. The effects of humic acid were observed mainly in combinations involving the sand only columns. The removal of HAdV in the sand column was improved from 92.7 % to 98.9 % due to the addition of humic acid. Unstable attachment of RV on sand media was observed; the plateau phase had a higher fluctuation with LRV ranging between 0.8 and 2.1. During DI flushing of the RV laden columns, we observed an RV peak concentration, corresponding to a LRV of 0.2, which increased the amount of virus release from 3.1 to 21.1 % (Table 6.2). The virus removal performances of the columns with hydrochar supplements were still high for both RV (99.8 %) and HAdV (99.9 %) in the presence of humic acid. The viral detachment in the DI flushing stage was comparable to the results from the experiments performed in the absence of humic acids addition, except for the combination RV and sand only.

6.4   Discussion
6.4.1   Effect of hydrochar supplement on virus removal in sand columns
The adsorptive behaviour of viruses may differ depending on the type of virus (even between different strains within the same species) due to their electrical charge and hydrophobicity differences (Schijven and Hassanizadeh, 2000). The majority of virus removal and transportation experiments have been based on the use of non-pathogenic model viruses such as MS2, PhiX 174 and PRD1. However, it was pointed out that the use of model viruses should be carefully considered due to the fact that pathogenic viruses could have different attachment/detachment and decay characteristics in the same aqueous conditions (Rigotto et al., 2011; Pang et al., 2014). In the same context, the use of active human pathogenic viruses employed in this research provides a more straightforward evaluation on virus removal. *Rotavirus WA* strain belongs to the G1 serotype which is recognized as one of the most common viral agents responsible for diarrheal diseases (Bahl et al., 2005; Santos and Hoshino, 2005). *Adenovirus C* type 2 was used as a representative strain in several studies for adenovirus in aqueous conditions (Yates et al., 2006; Baxter et al., 2007; Rigotto et al., 2011).

The virus removal during transport through porous media is a complex mechanism involving attractive (e.g. London-van der Waals and hydrophobic) and repulsive (e.g. electrostatic repulsion and steric) forces that may exist between two surfaces. These forces are mainly governed by solution chemistry such as pH, organic carbon content and ionic strength as well as by the properties of the virus and solid surfaces such as hydrophobicity, surface charge and specific surface area (Schijven and Hassanizadeh, 2000). The hydrochar supplement enhanced the removal of both RV and HAdV in all experimental conditions employed in this research. This improvement could be explained by the introduction of a larger surface area composed of hydrophobic meso-macroporous surfaces into the column media. Considering the size of RV and HAdV, which ranges between 70 and 100 nm (Friefeld et al., 1984; Stewart et al., 1993; Parashar et al., 1998), the presence of meso-macro porous (> 40 nm) surfaces would have provided more favourable attachment sites for viruses (Martin, 1980). This agrees well with previous research on the virus removal using activated carbon (Matsushita et al., 2013). The rough surfaces composed of mesopores facilitated the attachment of bacteriophage MS2 and Qß.

The negative zeta-potential value of hydrochar suggests repulsive electrostatic interactions between the hydrochar surface and the virus. The virus attachment to the hydrochar surface under these unfavourable conditions could perhaps be attributed to the hydrophobic surface of the hydrochar, thereby enforcing hydrophobic interactions between virus and hydrochar surface (Bales et al., 1991; Bales et al., 1993). Also, it was reported that viruses can attach to negatively charged surfaces owing to heterogeneous surface charge distribution. Because the zeta-potential only measures the net surface charge, the hydrochar surface might possess positive or less negative zones which favour virus attachment (Elimelech et al., 2000). The substantial virus removal in the sand columns without hydrochar supplements could be contributed to the same phenomenon. Foppen et al. (2006) reported that a considerable amount of PRD1 was removed in breakthrough experiments performed with columns packed by pre-washed quartz sand of 99.1% purity. The authors attributed this to microscopic chemical heterogeneities on the sand surface. This idea can be supported by previous studies carried out with glass beads with high purity (Li et al., 2004; Tufenkji and Elimelech, 2005).

6.4.2 Role of secondary energy minimum

The interaction between virus and column media can be explained by the Derjaguin-Landau-Verwey-Overbeek (DLVO) theory. While the viral attachment in the primary energy minimum (deep energy "well" close to the surface) is considered to be irreversible, the virions reside in the secondary energy minimum can be released (Kallay et al., 1987; Ryan and Gschwend, 1994). The flushing of DI water into virus-retained columns released both RV and HAdV into the effluent to a certain extent (Table 6.2). A similar phenomenon was observed in previous research investigating the retention of E. coli in column set-ups (Redman et al., 2004; Foppen et al., 2007; Chung et al., 2014). This detachment could be attributed to the decrease of ionic strength in the pore water which enforced electrostatic repulsion between the colloids and the collectors. The amount of virus release could be interpreted as a comparative indicator for virus residing in the secondary energy minimum. In this sense, among the experiments performed without humic acid in the influent, the role of the secondary energy minimum was more pronounced for RV than HAdV. Approximately

24% of the RV fed into hydrochar amended columns was released while the release from RV-sand, HAdV-sand and HAdV-hydrochar experiments was only 3.1 - 5.3 % (Table 6.2). The addition of humic acid into the influent increased the release of RV from the sand columns, which may be attributed to weakening of the hydrophobic attraction (Schijven and Hassanizadeh, 2000), whereby the dissolved organic matter disrupts the hydrophobic bonds between RV and the sand media, resulting in an increased release of RV during the DI flushing phase. On the other hand, we speculate that attachment of HAdV relied less upon hydrophobic attractions compared to RV.

Since RV and HAdV were tested under identical conditions, it is apparent that differences between two viruses induced different responses in DI water flushing and addition of humic acid to the feed suspension. The explanation for differences in attachment / detachment behaviour of RV and HAdV is still ambiguous due to a lack of relevant research with infectious RV and HAdV. Because RV and HAdV are similar in their size (70-100 nm) and morphology (icosahedral shape), the difference in surface characteristics would have resulted in major differences in attachment / detachment behaviour (Friefeld et al., 1984; Stewart et al., 1993; Parashar et al., 1998). Further quantitative assessments on hydrophobicity and surface charge of virus particles are therefore needed to improve the functioning of water treatment systems.

The role of secondary energy minimum could be of interest for adsorptive removal of contaminants from aqueous solutions because changes in influent characteristics (e.g. low ionic strength / high pH) might release pathogens accumulated on the adsorbent surface into the effluent. Since we observed that a considerable proportion of the virus particles resided in the secondary energy minimum, its application in water treatment should be carefully considered when large fluctuations in influent characteristics are expected.

6.4.3 Effect of humics in the feed solution

Despite the fact that humic substances have a negative surface charge and can compete with viruses for the same attachment sites (Charles, 1984), the humic acid addition into the influent did not reduce the virus removal performance in both sand and hydrochar amended columns. Moreover, the HAdV removal performance by the sand column was improved when humic acid was added to the influent. The adverse effect of humic acid on viral attachment may not always be apparent in case the adsorption sites are not fully saturated by humic substances (Gerba and Lance, 1978). In our research, the favourable attachment sites for negatively charged agents in the column media might have been enough for accommodating both humic acid and viruses dosed at relatively low concentrations.

6.4.4 (RT)-qPCR technology for virus quantification in water treatment

In microbial water quality analysis, PCR-based technologies have been regarded as the golden standard enabling sensitive quantitative and qualitative analysis. However, the main limitation of the (RT)-qPCR technology in evaluating water treatment technology is its inability to distinguish infectious and non-infectious microorganisms. Because the target nucleic acid could be persistent in environmental conditions, residual DNA or RNA from inactivated viruses can also give positive signals in the qPCR process, leading to an over-estimation of microbial risks (Josephson et al., 1993; Masters et al., 1994). Previous research compared the standard PFU method and a RT-qPCR assay for MS2 phage quantification in a slow sand filtration set-up. Here, the MS2 RNA was not degraded, and induced a gap between the results from two methods (Lodder et al., 2013Lodder et al., 2013). In the same context, the virus detected in the samples from breakthrough analyses might have been inactivated to a certain extent via disintegration of receptor and / or capsid during attachment-detachment processes (Harvey and Ryan 2004). The results from this research must be interpreted as a conservative indicator for the worst-case scenario. The viruses in the effluent samples from loading-deloading or the DI flushing stage could have been inactivated, but are still detected in (RT)-qPCR assessments.

Acknowledgements

This research was funded by the Korean Church of Brussels, Mangu Jeja Church (Seoul, Korea), and the Netherlands Ministry of Development Cooperation (DGIS) through the UNESCO-IHE Partnership Research Fund. It was carried out in the framework of the research project 'Addressing the Sanitation Crisis in Unsewered Slum Areas of African Mega-cities' (SCUSA).

6.5 References

Aertgeerts, R. (2009) Progress and challenges in water and sanitation. Desalination 248(1-3), 249-255.

Bahl, R., Ray, P., Subodh, S., Shambharkar, P., Saxena, M., Parashar, U., Gentsch, J., Glass, R., Bhan, M.K. and Delhi Rotavirus Study, G. (2005) Incidence of severe rotavirus diarrhea in New Delhi, India, and G and P types of the infecting rotavirus strains. Journal of Infectious Diseases 192, S114-S119.

Bales, R.C., Hinkle, S.R., Kroeger, T.W., Stocking, K. and Gerba, C.P. (1991) Bacteriophage adsorption during transport through porous media: chemical perturbations and reversibility. Environmental Science & Technology 25(12), 2088-2095.

Bales, R.C., Li, S., Maguire, K.M., Yahya, M.T. and Gerba, C.P. (1993) MS-2 and poliovirus transport in porous media: Hydrophobic effects and chemical perturbations. Water Resources Research 29(4), 957-963.

Baxter, C.S., Hofmann, R., Templeton, M.R., Brown, M. and Andrews, R.C. (2007) Inactivation of adenovirus types 2, 5, and 41 in drinking water by UV light, free chlorine, and mlonochlorarnine. Journal of Environmental Engineering-Asce 133(1), 95-103.

Boom, R., Sol, C., Beld, M., Weel, J., Goudsmit, J. and Wertheim-van Dillen, P. (1999) f alpha-casein to silica particles. Journal of Clinical Microbiology 37(3), 615-619.

Boom, R., Sol, C.J.A., Salimans, M.M.M., Jansen, C.L., Wertheimvandillen, P.M.E. and Vandernoordaa, J. (1990) Rapid and simple method for purification of nucleic-acids. Journal of Clinical Microbiology 28(3), 495-503.

Carter, M.J. (2005) Enterically infecting viruses: pathogenicity, transmission and significance for food and waterborne infection. Journal of Applied Microbiology 98(6), 1354-1380.

Chapron, C.D., Ballester, N.A., Fontaine, J.H., Frades, C.N. and Margolin, A.B. (2000) Detection of astroviruses, enteroviruses, and adenovirus types 40 and 41 in surface waters collected and evaluated by the information collection rule and an integrated cell culture-nested PCR procedure. Applied and Environmental Microbiology 66(6), 2520-2525.

Charles, G. (1984) Advances in Applied Microbiology. Allen, I.L. (ed), pp. 133-168, Academic Press.

Chung, J.W., Foppen, J.W., Izquierdo, M. and Lens, P.N.L. (2014) Removal of Escherichia coli from Saturated Sand Columns Supplemented with Hydrochar Produced from Maize. J. Environ. Qual. 0(0), -.

Crabtree, K.D., Gerba, C.P., Rose, J.B. and Haas, C.N. (1997) Waterborne adenovirus: A risk assessment. Water Science and Technology 35(11–12), 1-6.

Dai, L.C., Wu, B., Tan, F.R., He, M.X., Wang, W.G., Qin, H., Tang, X.Y., Zhu, Q.L., Pan, K. and Hu, Q.C. (2014) Engineered hydrochar composites for phosphorus removal/recovery: Lanthanum doped hydrochar prepared by hydrothermal carbonization of lanthanum pretreated rice straw. Bioresource Technology 161, 327-332.

Elaigwu, S.E., Rocher, V., Kyriakou, G. and Greenway, G.M. (2014) Removal of Pb2+ and Cd2+ from aqueous solution using chars from pyrolysis and microwave-assisted hydrothermal carbonization of Prosopis africana shell. Journal of Industrial and Engineering Chemistry 20(5), 3467-3473.

Elimelech, M., Nagai, M., Ko, C.H. and Ryan, J.N. (2000) Relative insignificance of mineral grain zeta potential to colloid transport in geochemically heterogeneous porous media. Environmental Science & Technology 34(11), 2143-2148.

Estes, M.K., Kang, G., Zeng, C.Q.Y., Crawford, S.E. and Ciarlet, M. (2008) Gastroenteritis Viruses, pp. 82-100, John Wiley & Sons, Ltd.

Foppen, J.W., van Herwerden, M. and Schijven, J. (2007) Transport of Escherichia coli in saturated porous media: Dual mode deposition and intra-population heterogeneity. Water Research 41(8), 1743-1753.

Foppen, J.W.A., Okletey, S. and Schijven, J.F. (2006) Effect of goethite coating and humic acid on the transport of bacteriophage PRD1 in columns of saturated sand. Journal of Contaminant Hydrology 85(3-4), 287-301.

Friefeld, B.R., Lichy, J.H., Field, J., Gronostajski, R.M., Guggenheimer, R.A., Krevolin, M.D., Nagata, K., Hurwitz, J. and Horwitz, M.S. (1984) The in vitro replication of adenovirus DNA. Current topics in microbiology and immunology 110, 221-255.

Funke, A. and Ziegler, F. (2010) Hydrothermal carbonization of biomass: A summary and discussion of chemical mechanisms for process engineering. Biofuels Bioproducts & Biorefining-Biofpr 4(2), 160-177.

Gerba, C.P. and Lance, J.C. (1978) Poliovirus removal from primary and secondary sewage effluent by soil filtration. Applied and Environmental Microbiology 36(2), 247-251.

Gomara, M.I., Simpson, R., Perault, A.M., Redpath, C., Lorgelly, P., Joshi, D., Mugford, M., Hughes, C.A., Dalrymple, J., Desselberger, U. and Gray, J. (2008) Structured surveillance of infantile gastroenteritis in East Anglia, UK: incidence of infection with common viral gastroenteric pathogens. Epidemiology and Infection 136(1), 23-33.

Hernroth, B.E., Conden-Hansson, A.C., Rehnstam-Holm, A.S., Girones, R. and Allard, A.K. (2002) Environmental factors influencing human viral pathogens and their potential indicator organisms in the blue mussel, Mytilus edulis: the first Scandinavian report. Applied and Environmental Microbiology 68(9), 4523-4533.

Josephson, K.L., Gerba, C.P. and Pepper, I.L. (1993) Polymerase chain-reaction detection of nonviable bacterial pathogens. Applied and Environmental Microbiology 59(10), 3513-3515.

Kallay, N., Barouch, E. and Matijević, E. (1987) Diffusional detachment of colloidal particles from solid/solution interfaces. Advances in Colloid and Interface Science 27(1–2), 1-42.

Kim, D., Lee, K. and Park, K.Y. (2014) Hydrothermal carbonization of anaerobically digested sludge for solid fuel production and energy recovery. Fuel 130, 120-125.

Kosek, M., Bern, C. and Guerrant, R.L. (2003) The global burden of diarrhoeal disease, as estimated from studies published between 1992 and 2000. Bulletin of the World Health Organization 81(3), 197-204.

Kumar, S., Loganathan, V.A., Gupta, R.B. and Barnett, M.O. (2011) An Assessment of U(VI) removal from groundwater using biochar produced from hydrothermal carbonization. Journal of Environmental Management 92(10), 2504-2512.

Li, X.Q., Scheibe, T.D. and Johnson, W.P. (2004) Apparent decreases in colloid deposition rate coefficients with distance of transport under unfavorable deposition conditions: A general phenomenon. Environmental Science & Technology 38(21), 5616-5625.

Libra, J.A., Ro, K.S., Kammann, C., Funke, A., Berge, N.D., Neubauer, Y., Titirici, M., Fuhner, C., Bens, O., Kern, J. and Emmerich, K. (2011) Hydrothermal carbonization of biomass residuals: A comparative review of the chemistry, processes and applications of wet and dry pyrolysis. Biofuels 2:89-124.

Lodder, W.J., van den Berg, H., Rutjes, S.A., Bouwknegt, M., Schijven, J.F. and Husman, A.M.D. (2013) Reduction of bacteriophage MS2 by filtration and irradiation determined by culture and quantitative real-time RT-PCR. Journal of Water and Health 11(2), 256-266.

Lutterodt, G., Basnet, M., Foppen, J.W.A. and Uhlenbrook, S. (2009) The effect of surface characteristics on the transport of multiple Escherichia coli isolates in large scale columns of quartz sand. Water Research 43(3), 595-604.

Martin, R.J. (1980) Activated Carbon Product Selection for Water and Wastewater Treatment. Industrial & Engineering Chemistry Product Research and Development 19(3), 435-441.

Masters, C.I., Shallcross, J.A. and Mackey, B.M. (1994) Effect of stress treatments on the detection of listeria-monocytogenes and enterotoxigenic escherichia-coli by the polymerase chain-reaction. Journal of Applied Bacteriology 77(1), 73-79.

Matsushita, T., Suzuki, H., Shirasaki, N., Matsui, Y. and Ohno, K. (2013) Adsorptive virus removal with super-powdered activated carbon. Separation and Purification Technology 107(0), 79-84.

Maunula, L., Klemola, P., Kauppinen, A., Soderberg, K., Nguyen, T., Pitkanen, T., Kaijalainen, S., Simonen, M.L., Miettinen, I.T., Lappalainen, M., Laine, J., Vuento, R., Kuusi, M. and Roivainen, M. (2009) Enteric Viruses in a Large Waterborne Outbreak of Acute Gastroenteritis in Finland. Food and environmental virology 1(1), 31-36.

Minani, J.M.V., Foppen, J.W. and Lens, P.N.L. (2014) Sorption of cadmium in columns of sand-supported hydrothermally carbonized particles. Water science and technology : a journal of the International Association on Water Pollution Research 69(12), 2504-2509.

Pang, L., Farkas, K., Bennett, G., Varsani, A., Easingwood, R., Tilley, R., Nowostawska, U. and Lin, S. (2014) Mimicking filtration and transport of rotavirus and adenovirus in sand media using DNA-labeled, protein-coated silica nanoparticles. Water Research 62, 167-179.

Pang, X.L.L., Lee, B., Boroumand, N., Leblanc, B., Preiksaitis, J.K. and Ip, C.C.Y. (2004) Increased detection of rotavirus using a real time reverse transcription-polymerase chain reaction (RT-PCR) assay in stool specimens from children with diarrhea. Journal of Medical Virology 72(3), 496-501.

Parashar, U.D., Bresee, J.S., Gentsch, J.R. and Glass, R.I. (1998) Rotavirus. Emerging Infectious Diseases 4(4), 561-570.

Parashar, U.D., Burton, A., Lanata, C., Boschi-Pinto, C., Shibuya, K., Steele, D., Birmingham, M. and Glass, R.I. (2009) Global Mortality Associated with Rotavirus Disease among Children in 2004. Journal of Infectious Diseases 200, S9-S15.

Parshetti, G.K., Chowdhury, S. and Balasubramanian, R. (2014) Hydrothermal conversion of urban food waste to chars for removal of textile dyes from contaminated waters. Bioresource Technology 161, 310-319.

Pina, S., Puig, M., Lucena, F., Jofre, J. and Girones, R. (1998) Viral pollution in the environment and in shellfish: Human adenovirus detection by PCR as an index of human viruses. Applied and Environmental Microbiology 64(9), 3376-3382.

Poerschmann, J., Weiner, B., Wedwitschka, H., Baskyr, I., Koehler, R. and Kopinke, F.D. (2014) Characterization of biocoals and dissolved organic matter phases obtained upon hydrothermal carbonization of brewer's spent grain. Bioresource Technology 164, 162-169.

Redman, J.A., Walker, S.L. and Elimelech, M. (2004) Bacterial adhesion and transport in porous media: Role of the secondary energy minimum. Environmental Science & Technology 38(6), 1777-1785.

Regmi, P., Moscoso, J.L.G., Kumar, S., Cao, X.Y., Mao, J.D. and Schafran, G. (2012) Removal of copper and cadmium from aqueous solution using switchgrass biochar produced via hydrothermal carbonization process. Journal of Environmental Management 109, 61-69.

Rigotto, C., Hanley, K., Rochelle, P.A., De Leon, R., Barardi, C.R.M. and Yates, M.V. (2011) Survival of Adenovirus Types 2 and 41 in Surface and Ground Waters Measured by a Plaque Assay. Environmental Science & Technology 45(9), 4145-4150.

Ryan, J.N. and Gschwend, P.M. (1994) EFFECTS OF IONIC-STRENGTH AND FLOW-RATE ON COLLOID RELEASE - RELATING KINETICS TO INTERSURFACE POTENTIAL-ENERGY. Journal of Colloid and Interface Science 164(1), 21-34.

Santos, N. and Hoshino, Y. (2005) Global distribution of rotavirus serotypes/genotypes and its implication for the development and implementation of an effective rotavirus vaccine. Reviews in Medical Virology 15(1), 29-56.

Schijven, J.F. and Hassanizadeh, S.M. (2000) Removal of viruses by soil passage: Overview of modeling, processes, and parameters. Critical Reviews in Environmental Science and Technology 30(1), 49-127.

Stewart, P.L., Fuller, S.D. and Burnett, R.M. (1993) Difference imaging of adenovirus - bridging the resolution gap between x-ray crystallography and electron-microscopy. Embo Journal 12(7), 2589-2599.

Sun, K., Ro, K., Guo, M., Novak, J., Mashayekhi, H. and Xing, B. (2011) Sorption of bisphenol A, 17α-ethinyl estradiol and phenanthrene on thermally and hydrothermally produced biochars. Bioresource Technology 102(10), 5757-5763.

Tufenkji, N. and Elimelech, M. (2005) Spatial distributions of Cryptosporidium oocysts in porous media: Evidence for dual mode deposition. Environmental Science & Technology 39(10), 3620-3629.

WHO and UNICEF (2010) Progress on Sanitation and Drinking-water: 2010 Update. WHO Press, Geneva, Switzerland.

Xue, B., Jin, M., Yang, D., Guo, X., Chen, Z., Shen, Z., Wang, X., Qiu, Z., Wang, J., Zhang, B. and Li, J. (2013) Effects of chlorine and chlorine dioxide on human rotavirus infectivity and genome stability. Water Research 47(10), 3329-3338.

Yates, M.V., Malley, J., Rochelle, P. and Hoffman, R. (2006) Effect of adenovirus resistance on UV disinfection requirements: A report on the state of adenovirus science. Journal American Water Works Association 98(6), 93-106.

Chapter 7: Simultaneous removal of rotavirus and adenovirus from artificial ground water using hydrochar derived from swine faeces

This chapter is based on:

Chung, J.W., Foppen, J.W., Breulmann, M., Clemens, A., Fuhner, C. and Lens, P. (submitted) Simultaneous removal of rotavirus and adenovirus from artificial ground water using sand column supplemented with hydrochar derived from swine faeces.

Abstract

Swine faeces was hydrothermally treated at two conditions: at 230 °C for 7 hr and 180 °C for 2 hr. The resulting solid products (hydrochar) were evaluated as virus adsorbents in water treatment. Simultaneous removal of pathogenic rotavirus (RV) and human adenovirus (HAdV) was investigated using a simple sand column set-up of 10 cm bed height with and without hydrochar supplement (1.5 %, *w/w*) at three flow rates. The removal efficiency of both viruses obtained from the sand-only column was 1.3 log at a flow rate 1 mL / min. As the flow rate increased, the removal efficiency decreased to ~ 0.6 log for rotavirus and ~ 0.9 log for adenovirus. In contrast, the removal efficiency of both viruses in a hydrochar-amended column was > 3 log (complete removal), and 2.1 log - > 3 log at elevated flow rates. Regardless of the type, hydrochar amendments in the sand columns significantly increased the virus removal performance at all applied flow rates. The amount of virus released in deionized (DI) water when flushed into the virus-retaining columns indicated that the secondary energy minimum played a more important role in the RV retention than that of HAdV. Zeta-potential and hydrophobicity measurements on hydrochar materials indicated that the improved virus removal performance of hydrochar-amended columns was induced by the provision of extra hydrophobic surfaces. This study provides evidence that faecal waste-derived hydrochar can be used as a competent virus adsorbent.

7.1 Introduction

Provision of adequate sanitation and clean water is an important challenge for public health in many developing countries. In spite of international efforts and notable progress in the last decades, still over 700 million people are disconnected from improved drinking-water sources, and 2.5 billion do not have an access to proper sanitation (WHO and UNICEF, 2014). In urban areas of developing countries, where sewer systems and sewer-based faecal waste treatment systems are not affordable, on-site sanitation technologies are predominantly used generating large quantities of untreated faecal waste which cause extensive environmental contamination (Koné, 2010).

With proper measures, however, faecal waste can be utilized as a valuable resource. Recently, Diener et al. (2014) suggested the conversion of faecal sludge into market products: dry sludge as fuel for combustion, animal protein (food for larvae of black soldier fly), feed stock for biogas generation, soil conditioner and building materials. In this context, hydrothermal carbonization (HTC), also known as wet pyrolysis, of faecal waste can be an attractive technology providing not only a better hygienic environment, but also valuable carbonaceous materials (Katukiza et al., 2012). Faecal waste, the most important source of human pathogens, is totally sanitized during the HTC process under wet and pressurized (~20 bar) conditions at relatively low temperatures (~200 °C). The HTC product is a thick slurry consisting of a solid particles (hydrochar) and liquid (process water) phase. The hydrochar can be utilized as pathogen-free agricultural supplements (Berge et al., 2013), energy source (Mumme et al., 2011), phosphorus source (Heilmann et al., 2014) and adsorbent in water and wastewater treatment for biotic and / or abiotic contaminant removal (Sun et al., 2011; Chung et al., 2014; Minani et al., 2014; Chung et al., 2015).

One of the main threats induced from improperly managed faecal waste is microbial contamination in surface water and groundwater of peri-urban areas (Katukiza et al., 2013). Similar to the sanitation issue, conventional centralized water treatment systems are inappropriate in places in need due to technical and financial limitations (Bartram et al., 2005). Instead, simple and low-cost decentralized (point-of-use) technologies such as a biosand filter, a ceramic filter, solar disinfection and chlorination (combination with flocculation) have been recommended as affordable solutions for safe drinking-water provision (Sobsey et al., 2008).

These technologies have, however, relatively limited removal capacities for the pathogenic viruses (Sobsey et al., 2008). Large quantities of highly infective enteric viruses are shed in faeces of infected persons: 10^5 - 10^{11} virons / g faeces of infected person (Farthing, 1989; Yezli and Otter, 2011). They are considered to be responsible for a number of outbreaks of waterborne diseases (Bosch, 1998). In this research, we investigated the potential use of faecal waste-derived hydrochar as an adsorbent for virus removal with potential application in water treatment. As a model of human excreta, fresh swine faeces was hydrothermally converted into carbonaceous adsorbents. Simultaneous removal of infective pathogenic rotavirus and adenovirus, two of the most significant viral agents responsible for diarrheal diseases was carried out by performing breakthrough analyses using a simple sand filtration set-up supplemented with a small amount of hydrochar. Also, the effect of the flow rate and the secondary energy minimum on the viral retention was investigated.

7.2 Methods and materials

7.2.1 Hydrochar

Fresh swine faeces was collected at the research farm of the Faculty of Veterinary Medicine of Leipzig University (Germany). The feedstock for HTC was prepared by mixing fresh swine faeces and deionized (DI) water at a dry matter content of 15.8 %. The pH was adjusted to 4.6 by addition of 0.5 M H_2SO_4. The hydrothermal conversion of swine waste was carried out in a high-pressure reactor (BR-300, Berghof, Eningen, Germany) equipped with a stirrer drive (BRM-1, Berghof, Eningen, Germany). Approximately, 55 g dry weight of swine manure was loaded in a 500 mL volume stainless steel vessel with polytetrafluoroethylene (PTFE) insert. Two experimental conditions were employed for hydrochar production: at 230 °C for 7 h (230-HTC) and at 180 °C for 2 h (180-HTC), representing two extremes of reaction severity (Funke and Ziegler, 2010). The reactions were performed under a heating rate of 2 K / min with a stirring speed of 150 rpm. The pressure and temperature inside the reactor were monitored by a data logger (BTC-3000, Berghof, Eningen, Germany). Afterwards, the reactor was cooled down to room temperature. Then, the gaseous product was released into an inverse volumetric cylinder immersed in water in order to measure the volume of gaseous products. The hydrochar was separated from the resulting slurry by vacuum filtration using a ceramic funnel and filter paper. The hydrochar was dried at 70 °C overnight and stored at room temperature. The weight of the resulting material at each step was measured. The moisture, volatile organic matter, fixed carbon and ash contents of oven-dried (at 70 °C) swine manure and both types of hydrochar were analyzed according to a standard protocol (ASTM, 2003).

7.2.2 Virus suspension

Active human pathogenic *Rotavirus WA* (RV) and *Human adenovirus C* type 2 (HAdV) were kindly provided by The Netherlands National Institute for Public Health and Environment (RIVM, Bilthoven, The Netherlands). Both are non-enveloped viruses and similar in their morphology (icosahedral shape) and size (80-100 nm) (Stewart et al., 1993; Parashar et al., 1998). Stocks of RV with a concentration of $10^{5.8}$ tissue culture infectious dose 50 ($TCID_{50}$) / mL and HAdV stocks with a concentration of 10^6 plaque - forming unit (PFU) / mL were aliquoted in Eppendorf vials and stored in a freezer (-80 °C).

The virus removal experiments were carried out using artificial ground water (AGW), which was prepared by dissolving 526 mg $CaCl_2.2H_2O$, 184 mg $MgSO_4.7H_2O$, 8.5 mg KH_2PO_4, 21.8 mg K_2HPO_4 and 17.7 mg Na_2HPO_4 in 1 L DI water (Chung et al., 2014). The pH and electrical conductivity (EC) of the AGW were ~ 6.8 and 1000 μS / cm, respectively. The influent was prepared by suspending RV and HAdV aliquots in AGW at a concentration of ~10^3 $TCID_{50}$ (RV) and PFU (HAdV) / mL. This influent was placed on a lab bench for 30 min prior to the virus removal experiments for stabilization of viruses in AGW. All operations with pathogenic viruses were carried out in a biosafety cabinet satisfying biosafety level (BSL) 2. Virus-contaminated materials were disinfected by autoclaving at 121 °C for 20 min.

7.2.3 Virus quantification

The nucleic acid of the virus was extracted by using chaotrophic buffers and silica colloids as previously described (Boom et al., 1990; Boom et al., 1999). Established reverse transcription-polymerase chain reaction (RT-qPCR) protocols were used to determine the concentration of RV (Chung et al., 2013) and HAdV (Chung et al., 2015) with minor modifications. Briefly, denaturation and annealing of RV RNA with random primers were performed in a reaction mix consisting of 5 μL nucleic acid extract, 0.3 μL random hexamer (Fermentas) and 8.7 μL DEPC treated water at 70°C for 5 min. Reverse transcription was carried out by addition of 0.3 μL (60 unit) RevertAid reverse transcriptase (Fermentas), 1 μL dNTP mix (4 mM of each dNTP, GenScript), and 5 μL 5x RT buffer and 4 μL DEPC treated water into the product of the previous step. Then, the resulting reverse transcription mix with a final volume of 25 μL was exposed at 25 °C for 10 min followed by at 42 °C for 60 min, and finally at 70 °C for 10 min. All thermal treatment in this research was carried out in a thermal cycler BioRad MJMini™ (real time PCR system, Miniopticon).

The real-time PCR assessments of RV cDNA or HAdV DNA were carried out using an identical protocol except for the use of primers and probes for RV (Pang et al., 2004) and HAdV (Hernroth et al., 2002). These probe and primers were synthesized by Biolegio (Nijmegen, The Netherlands). Briefly, 4 μL of template (RV cDNA or nucleic acid extract) was added to a qPCR mix containing 2.5 μL home-made qPCR buffer (150 mM Tris-HCl pH 8.2 at 25°C, 300 mM KCl, 50 mM $(NH_4)_2SO_4$, 25 mM $MgCl_2$, and 0.2% BSA), 1 μl dNTP mix, 1 μL (0.5 unit) Taq polymerase (GenScript), 15.4 μL DEPC treated water, 0.4 μL of forward and reverse primer (both at a 200 nM final concentration), and 0.3 μL probe (150 nM final concentration). The thermal cycling consisted of an initial denaturation at 95 °C for 5 min, and 40 cycles of denaturation at 94 °C for 20 s followed by annealing / extension at 60 °C for 60 s. The fluorescence signal was monitored at the end of every cycle. The threshold cycle values (Cts) obtained from qPCR of samples were converted into actual virus concentrations using regression curves derived from 10-fold dilution series of standard virus stocks (data not shown).

7.2.4   Material characterization

Zeta potential

The zeta potential values of the two types of hydrochar were determined in the pH range from 4 to 10 using a Zetasizer Nano ZS (Malvern, UK) equipped with a MPT-2 pH auto-titration unit. The hydrochar samples were prepared by repetitive washing in AGW prior to the zeta potential measurements. The concentration of hydrochar in the sample was controlled to have an adequate attenuator (6 - 8) selection of the instrument.

Hydrophobicity

To analyze hydrophobicity of two types of hydrochar samples, static contact angle analysis was performed using a Drop Shape Analyzer (DSA100, KRÜSS, Hamburg, Germany). A flat surface of hydrochar was prepared by pelletizing of powdered hydrochar (Jeong et al., 2009) using a manual hydraulic press (Atlas Manual Hydraulic Press, Kent, UK).

7.2.5   Column experiments

Column preparation

The virus removal efficiencies of both hydrochars (230-HTC and 180-HTC) were investigated by performing breakthrough analyses in a simple sand filtration set-up as described previously (Chung et al., 2014). Briefly, acid-washed quartz sand with 99.1 % purity (Kristall quartz-sand, Dorsilit, Germany) was used as a supporting matrix for hydrochar particles. A cumulative mass distribution curve was plotted through sieve analysis of sand granules; a D_{50} value of 0.79 mm and a U value (D_{10} / D_{60}) of 1.81 were obtained (Matthess et al., 1991). A borosilicate glass column with a 2.5 cm inner diameter was packed with either sand or with a sand-hydrochar mixture (at 1.5 %, dry weight hydrochar / sand) to have a 10 cm bed height (Omnifit, Cambridge, U.K.). In order to prevent air entrapment and channeling in the sand-hydrochar matrix, the column was regularly agitated and the packing materials were carefully compacted by using a glass rod during the packing process. Then, the column was connected to a peristaltic pump (MasterFlex model 77201-60, Vernon Hills, Illinois, USA) and washed with DI water overnight to remove residual fine particles and chemicals from the column matrix. Prior to the virus breakthrough analysis, the column was saturated with AGW.

Design of the virus removal experiments

To investigate the simultaneous removal of RV and HAdV, both types of viruses were seeded in the influent AGW. The breakthrough analysis was carried out by flushing of 50 mL virus-containing AGW (loading) followed by flushing 50 mL virus-free AGW (deloading), and finally 100 mL DI water was fed into the column. It was anticipated that the flushing with DI water can be an indicator for the role of the secondary energy minimum in the viral retention in the column. The low ionic strength of DI water (4 - 5 µS / cm) would have increased the repulsive electrostatic force between the virus and column media surface by expansion of the electrical double layer between the virus and the surface of column media. As a result, a certain amount of viruses deposited in the secondary energy minimum will be released (Foppen et al., 2007; Chung et al., 2014).

The effect of the flow rate in virus retention and release behavior was examined by applying 3 flow rates in upward direction: 1, 2.5 and 5 mL / min, which correspond to 0.12, 0.3, and 0.6 m / h, respectively. These can be considered as a high range of slow sand filtration rates. It was assumed that the lower flow rate induced a longer retention time and weaker shear stress, and the opposite conditions were true for the higher flow rates. In total 36 breakthrough curves (BTC) were obtained from 18 duplicate experiments with 2 viruses (RV and HAdV), 3 packing materials (sand, sand-230-HTC or sand-180-HTC) and 3 flow rates (1, 2.5 and 5 mL / min). The concentration of RV and HAdV in the effluent was measured at 5 min intervals by (RT)-qPCR assays as described above. Every breakthrough experiment was performed in duplicate.

7.3 Results
7.3.1 Hydrothermal carbonization of swine waste
Swine faeces was hydrothermally converted at two reaction conditions: 230 °C for 7 h (230-HTC, representing severe reaction) and 180 °C for 2 h (180-HTC, representing mild reaction). The profiles of pressure and temperature development inside the reactor are given in Figure 1. The operating pressure during the HTC process maintained at 35 - 38 bar for the reaction at 230 °C and 10 - 13 bar for the reaction at 180 °C. These different hydrothermal conditions resulted in different characteristics of the hydrochar materials. The solid product yield, gas production, and the constitution of materials are summarized in Table 1. The yield of solid product was 14.1 % higher for 180-HTC than 230-HTC. In contrast, the volume of gaseous product was twice more with 230-HTC than with 180-HTC. The total amount of carbonaceous contents in solid products decreased during both HTC processes. It was apparent that a certain amount of volatile organic matter in the feedstock was converted into different forms such as fixed carbon in hydrochar, soluble products in process water or gaseous products.

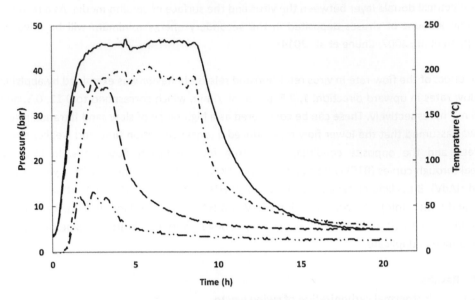

Figure 1: Temperature and pressure profile during HTC of swine waste.
Legends: (————) 230-HTC temperature, (——) 180-HTC temperature, (—·—) 230-HTC
pressure and (—··) 180-HTC pressure.

Table 1. Composition of HTC materials, solid yield and gas production capacity derived from
swine manure

Sample	Moisture (%)	Volatile organic matter (%)	Fixed carbon (%)	Total carbon (%)	Ash (%)	Solid yield (%)	Gas (mL / g dry manure)
Dry swine manure	[a] 8.4	62.6	13.5	76.1	15.5	[c] n. a.	n. a.
230-HTC	1.9 [b] (1.1)	44.2 (26.1)	38.2 (22.5)	82.4 (48.6)	15.7 (9.26)	[d] 59.0	[e] 46.0
180-HTC	1.0 (0.7)	32.4 (23.7)	50.9 (37.2)	83.3 (60.9)	15.6 (11.4)	73.1	22.1

[a] average % of duplicate measurements

[b] the number in brackets indicates the mass conversion of each content (solid yield ×
average %)

[c] not available

[d] average % of triplicate measurements, dry weight hydrochar / dry weight feedstock

[e] average of triplicate measurements, mL gaseous product / dry weight feedstock

7.3.2 Material characterization

Zeta potential

The zeta potential values of 230-HTC and 180-HTC were all negative in the pH range from 4 to 10 (Figure 2). This observation implied that slightly stronger electrostatic repulsion existed in the columns with 230-HTC (-16 mV) than the ones with 180-HTC (-13 mV) at the operational pH of AGW (pH 6.8).

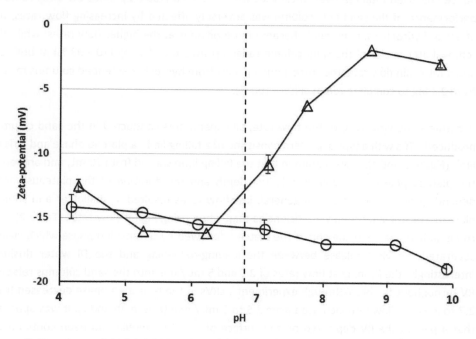

Figure 2: Zeta potential of (o) 230-HTC and (Δ) 180-HTC as a function of pH. The solid lines represent the average zeta-potential value of triplicate assessments and error bars indicate the standard deviation. The dashed line indicates the pH value of AGW (6.8).

Hydrophobicity

Comparable contact angle results were obtained from both hydrochar discs: 103.8 (± 1.2) ° for HTC-180 (average of results from four replications ± standard deviation) compared to 101.8 (± 2.6) ° for HTC-230. These values indicated that hydrochar was strongly hydrophobic.

7.3.3 Column experiments

The BTCs of HAdV and RV are given in Figures 3 and 4, respectively. The virus concentration in the effluent was expressed as a log removal value (-log C / C_0, LRV). Virus removal efficiencies were calculated from the effluent samples collected in the loading (5-50 mL) and deloading (55-100 mL) phase. The amount of viruses observed in the DI water flushing (105-200 mL) stage was considered as an indicator for the role of the secondary energy minimum.

Amendments with either 180-HTC or 230-HTC in the sand media had clearly enhanced the virus removal performance at all flow rates employed. In terms of virus removal efficiency, both hydrochar supplements did not show clear differences. Complete virus removal was achieved from both hydrochar-amended columns at a 1 mL / min flow rate (Figure 3a and Figure 4a). At elevated flow rates, only irregular release of viruses at low concentrations (LRV >2) was observed in the loading-deloading phase (Figure 3b, c, Figure 4b and c), and hence improved total removal efficiencies were obtained (Table 2). Since the virus removal performance of the sand-only column was adversely affected by increasing flow rates, the effect of hydrochar amendments became more obvious at the higher flow rates: while the removal efficiencies of the sand column ranged from 0.6 to 1.2 log (70 - 93 %) at both 2.5 and 5 mL / min flow rate, the corresponding ones from hydrochar amended columns ranged from 2.1 log to complete removal (99 - 100 %).

Regardless the virus type or the flow rate, all experiments conducted in the sand columns produced BTCs with a typical pattern consisting of a falling limb, a plateau phase and a rising limb (Figure 3 and 4). It was common that the falling limb started from 10 mL, and arrived to the plateau phase at around 20 mL. The depth and the duration of the plateaus varied depending on the flow rates. In general, low flow rates resulted in shallower and shorter plateaus, or in higher removal efficiencies than when using higher flow rates (Table 2). The rising limb appeared at the end of the plateau, followed by a transition phase which clearly separated the virus release between the loading-deloading and the DI water flushing. Interestingly, the flushing at flow rates of 2.5 and 5 mL / min into the sand columns released RV throughout the breakthrough experiments. LRVs in the transition phase decreased from 2.7 to 2 as the flow rate increased from 2.5 to 5 mL / min (Figure 4b and c). It was apparent that a part of the RV deposited on the surface of the sand media had been continuously detached. This phenomenon was not observed in HAdV BTCs obtained from sand-only columns.

In general, the viral detachment upon flushing with DI water was larger for RV than HAdV for all experimental conditions employed (Table 2). For each type of virus, varying the flow rate did not significantly alter the amount of virus released. Though it could be speculated that the role of secondary energy minima in the RV retention by the sand column was more significant at 5 mL / min than the lower flow rates, the result should be carefully interpreted because of the possible effect of increased physical stress that might have facilitated the viral detachment, shown as a decrease in LRV during the transition phase (Figure 4c). This could lead to overestimating the RV residing in the secondary energy minimum.

Table 2. Removal and release of viruses in the breakthrough analysis with deionized water flushing.

Flow rate (mL / min)		Sand		180-HTC+Sand		230-HTC+Sand	
		RV	HAdV	RV	HAdV	RV	HAdV
1.0	LRV	[a] 1.3	1.3	> 3.0	> 3.1	> 3.0	> 3.1
	Release in DI water (%)	[b] (4.0)	(1.6)	(13.0)	(3.1)	(9.9)	(1.5)
2.5	LRV	0.8	1.2	2.5	2.6	> 3.0	2.2
	Release in DI water (%)	(3.3)	(2.3)	(3.8)	(3.0)	(11.5)	(2.4)
5.0	LRV	0.6	0.9	2.5	2.4	2.1	2.5
	Release in DI water (%)	(11.4)	(3.9)	(14.6)	(7.1)	(8.3)	(2.6)

[a] Average log removal efficiency obtained from duplicated breakthrough analysis. The values with a sign of inequality represent C_0, which indicates complete virus removal.
[b] Average percentage of virus release in deionized water flushing stage.

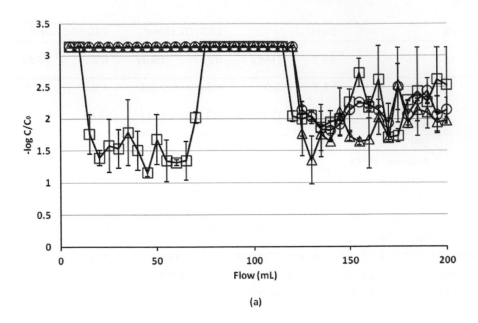

(a)

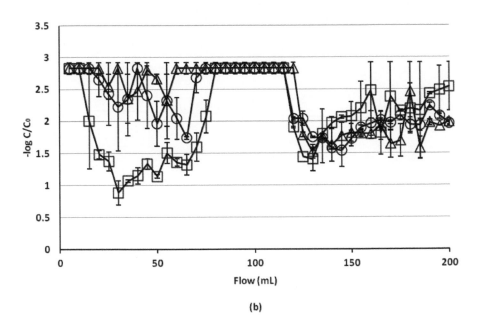

(b)

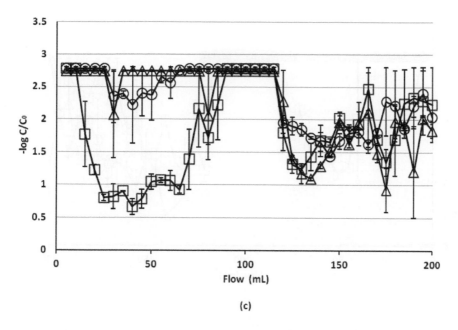

(c)

Figure 3: Adenovirus breakthrough curves at a flow rate of (a) 1 mL / min, (b) 2.5 mL / min and (c) 5 mL / min: Solid lines represent the mean log removal values (-log C/C_0, LRV) and error bars indicate the variation of the two breakthrough analyses. Note that the deionized water flushing started at 100 min. The log removal values of samples under the detection limit were presented as C_0. Columns: (□) Sand, (Δ) 180-HTC and (○) 230-HTC. Note that the DI water flushing started at 100 mL.

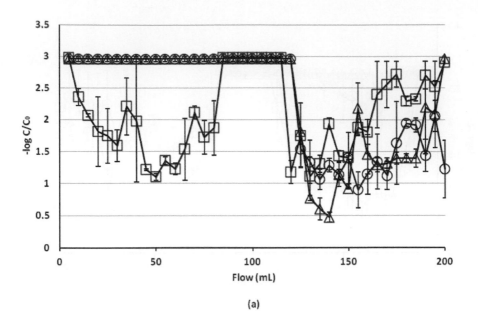

(a)

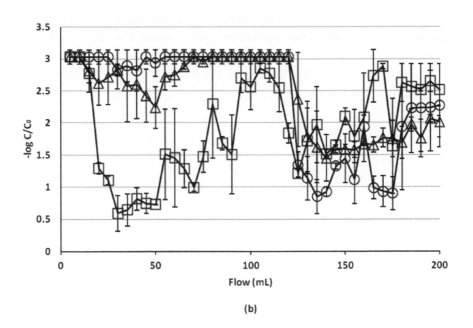

(b)

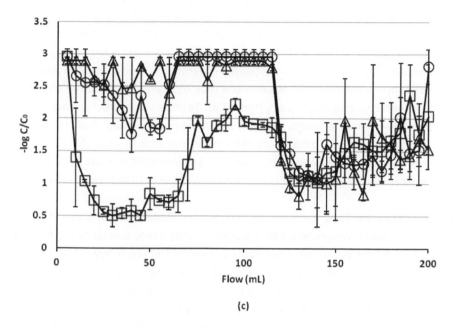

(c)

Figure 4: Rotavirus breakthrough curves at a flow rate of (a) 1 mL / min, (b) 2.5 mL / min and (c) 5 mL / min: Solid lines represent the mean log removal values (-log C/C_0, LRV) and error bars indicate the variation of the two breakthrough analyses. Note that the deionized water flushing started at 100 min. The log removal values of samples under the detection limit were presented as C_0. Columns: (○) 230-HTC, (Δ) 180-HTC and (□) Sand. Note that the DI water flushing started at 100 mL.

7.4 Discussion

7.4.1 Improved virus removal with hydrochar supplement in sand columns

This study showed that the virus removal efficiency of sand columns was significantly increased by supplementing hydrochar (1.5 %, w/w) into the column media. It shows that manure derived hydrochar is as effective in virus removal from AGW as sewage sludge derived hydrochar (Chung et al., 2015) when supplemented in sand columns.

The adsorptive retention capacity of sand columns (with or without hydrochar supplement) can be proportional to the surface area and the favorable surface functionality of the column media. The viral attachment on solid media is primarily controlled by van der Waals, electrostatic (electrical double-layer) and hydrophobic interactions that vary depending on surface characteristics of the virus and media, temperature and solution chemistry (Israelachvili and Wennerstrom, 1996; John and Rose, 2005). Since the net surface charge of non-enveloped viruses is determined by functional groups on the viral capsid and their protonation-deprotonation states, the solution pH plays a key role determining the electrostatic interaction (Michen and Graule, 2010). Most frequently, viruses in polar media (such as water) have a negative surface charge at a neutral pH (Michen and Graule, 2010). Because both hydrochar types (180-HTC and 230-HTC) in the AGW also showed a negative zeta-potential (Figure 2), repulsive electrostatic interactions are expected to be dominant between viruses and media surfaces in the column experiments. Therefore, improvement of virus removal in hydrochar-amended columns can thus be mainly attributed to an increase in surface area with strong hydrophobicity (see 3.2.2). Because the hydrochar powders were much smaller than sand granules, hydrochar supplements would have provided a meaningful extra surface area onto which the viruses can attach. Also, the results from hydrophobicity measurements on hydrochar samples suggested an increase in hydrophobic attraction between viruses and the hydrochar surface (Bales et al., 1991; Bales et al., 1993).

Another mechanism that likely contributed to the RV and HAdB removal is viral deposition in flow stagnation zones: viruses would be retained without direct contact to the hydrochar surface, but immobilized by the hydrochar surface owing to the secondary energy minimum (Table 2) (Redman et al., 2004; Brow et al., 2005). Other possible causes for virus deposition under unfavorable conditions include heterogeneous charge distribution, the presence of localized patches with positive charge that would have induced attractive electrostatic interaction despite a net negative charge (Elimelech and Omelia, 1990; Elimelech et al., 2000).

Reversible attachment of virus particles is of major interest for effluent quality surveillance, because changes in influent ionic strength and / or pH can result in undesirable releases of infectious virus particles accumulated on the hydrochar surface. In our previous work carried out in a comparable experimental set-up using sand columns supplemented with a sewage sludge-derived hydrochar (1.5 %, *w/w*), the release of HAdV in DI water flushing was comparable to the results of this research (~5 %) (Chung et al., 2015). In contrast, the RV release from sewage sludge-amended sand columns was greater (24 %, Chung et al., 2015) than the release from 180-HTC (9 %, Table 2) or 230-HTC (13 %, Table 2). It can be speculated that the faecal waste derived-hydrochar retained a larger proportion of RV in the primary energy minimum, which is considered as irreversible attachment. From this viewpoint, faecal waste-derived hydrochar seems better adsorbents than the sewage sludge-derived hydrochar.

The virus concentrations in the samples were measured by PCR-based methods. Due to their inability in assessing viral infectivity, the virus removal performance observed in this research must be interpreted as the worst-case scenario assuming all viruses in the samples are still infective (Lodder et al., 2013).

7.4.2 Effect of flow rate and secondary energy minimum

Colloidal deposition is affected by the flow rate (Li et al., 2005; Johnson et al., 2007). Depending on the presence or absence of an energy barrier to deposition, the flow rate affects colloidal deposition and release (re-entrainment) in opposite ways: in the presence of an energy barrier, colloidal deposition rates decrease and detachment rates increase with increasing flow rate (Tong and Johnson, 2006; Johnson et al., 2007). Since both biological colloids and media surfaces (silicate mineral grains) carry an overall negative surface charge (unfavorable condition) under most environmental conditions (Davis, 1982; Tipping and Cooke, 1982), an increase in flow rate has a negative impact on pathogen removal in adsorption-based water treatment systems. Our observations agree well with these investigations: the removal efficiency decreased with increasing flow rate (Table 2).

The effect of increased flow rate (flow velocity) on the deposition and release behavior was attributed to an increase in fluid drag (Tong and Johnson, 2006; Johnson et al., 2007). Li et al (2005) suggested a close relation between the effects of the fluid drag and the role of the secondary energy minimum on colloidal deposition and re-entrainment. Briefly, the increased fluid drag may, to a certain extent, shift the deposition of colloids from the primary energy minimum (strong attachment) to the secondary energy minimum (weak attachment). Then, this increase in the role of the secondary energy minimum would have exposed more colloids to chances of detachments. Potential mechanisms for the effect of fluid drag on deposition and re-entrainment behavior of colloids can be summarized as follows (Li et al., 2005): (i) decrease in stagnation flow zones in which colloids can accommodate, (ii) increase in colloidal diffusion induced by enlarged concentration gradients between fluid and zones in which colloids accumulate, (iii) increased hydrodynamic collisions between mobile and deposited colloids, and (iv) increased hydrodynamic torque relative to the adhesive torque. The effect of increased flow rates was more prominent in sand-only columns than in hydrochar-amended columns (Table 2). We speculated that hydrophobic interaction between hydrochar surfaces and virus particles had provided an additional attractive force which compensated the negative impact on the virus removal induced from increased fluid drag.

The results of this study suggest that the secondary energy minimum was more pronounced for RV than HAdV. For each experimental condition, regardless of the type of column packing material and flow rate, the RV release in DI water flushing was greater than the release of HAdV (Table 2). This observation corresponded well with our previous work (Chung et al., 2015). The higher variation of the RV removal efficiency depending on the flow rate in the sand media can be explained by the larger role of the secondary energy minimum (weak attachment), suggesting larger detachment rates throughout flushing experiments. The extended tailing of RV and elevated LRV level in the intermediate phase in the sand-only columns at 5 mL / min (Figure 3c) support this idea (Li et al., 2005). The response to the flow rate variation can be considered as virus-specific. Despite the similarities of both viruses in their morphology (Friefeld et al., 1984; Stewart et al., 1993; Parashar et al., 1998), the difference in surface characteristics can explain the different response to the flow rate increase (Schijven and Hassanizadeh, 2000).

7.4.3   HTC for faecal waste treatment

The difference in reaction intensity during the hydrochar production resulted in a different hydrochar yield and composition (Figure 1 and Table 1). However, no clear difference in virus removal efficiency between 230-HTC and 180-HTC was observed for all experiments (Table 2). If we limit the potential use of hydrochar to a virus adsorbent in groundwater treatment, HTC produced at 180 °C could be a more attractive option due to the higher hydrochar yield (Table 1) and lower energy requirement. Higher temperatures and extended reaction times in HTC production at 230 °C yield more operational costs. Also, the higher pressure induced in 230-HTC might necessitate a more pressure-resistant reactor, which will increase the capital costs.

Sustainable implementation of faecal waste treatment with HTC would be feasible in case the cost needed for HTC is lower than current treatment in specific cases (e.g. planted or unplanted sludge drying bed) or when the cost can be compensated by the benefits from HTC products (Koné, 2010). Since the economic value of the end-products varies significantly depending on implementation sites and their market demands (Diener et al., 2014), more research on faecal waste-HTC products is needed (e.g. calorific value, plant-available nutrient and biogas production capacity). Also, indirect benefits such as better hygienic surveillance (Hutton et al., 2007) or preventing deforestation (replacement of conventional plant-based energy source with HTC products) (Miles and Kapos, 2008) need to be considered as an asset.

Since the characteristics of hydrochar are usually determined by the nature of the feed stock and several process parameters such as hydrous conditions, temperature, residence time, pressure, solid load and pH (Funke and Ziegler, 2010), optimization of the HTC process (e.g. co-carbonization of faecal waste with locally available agricultural residues) pursuing maximum economic value of the HTC products is an important topic for further research.

7.5 Conclusions

This study showed that:

i) Hydrochar derived from different HTC conditions (180-HTC and 230-HTC) showed similar efficient removal of RV and HAdV for all flow rates employed. This can be mainly attributed to the provision of extra hydrophobic surfaces enforcing hydrophobic attraction between the viruses and column media.

ii) In sand-only columns, an increase in flow rate decreased the virus removal efficiency. Supplementing the columns with hydrochar mitigated the negative impacts from the increased flow rate in the virus retention.

iii) More release was observed for RV than HAdV during DI water flushing, despite the similar size and shape of both viruses. Apparently, the contribution of the secondary energy minimum in the virus retention was more determined by surface characteristics of the viruses than their size and shape.

Acknowledgements

This research was funded by the Korean Church of Brussels, Mangu Jeja Church (Seoul, Korea), and the Netherlands Ministry of Development Cooperation (DGIS) through the UNESCO-IHE Partnership Research Fund. It was carried out in the framework of the research project 'Addressing the Sanitation Crisis in Unsewered Slum Areas of African Mega-cities' (SCUSA).

7.6 References

ASTM (2003) Standard Test Method for Chemical Analysis of Wood Charcoal, ASTM D1762-84, ASTM International, West Conshohocken, PA, 2013.

Bales, R.C., Hinkle, S.R., Kroeger, T.W., Stocking, K. and Gerba, C.P. (1991) Bacteriophage adsorption during transport through porous media: chemical perturbations and reversibility. Environmental Science & Technology 25(12), 2088-2095.

Bales, R.C., Li, S., Maguire, K.M., Yahya, M.T. and Gerba, C.P. (1993) MS-2 and poliovirus transport in porous media: Hydrophobic effects and chemical perturbations. Water Resources Research 29(4), 957-963.

Bartram, J., Lewis, K., Lenton, R. and Wright, A. (2005) Focusing on improved water and sanitation for health. Lancet 365(9461), 810-812.

Berge, N.D., Kammann, C., Ro, K. and Libra, J. (2013) Sustainable Carbon Materials from Hydrothermal Processes, pp. 295-340, John Wiley & Sons, Ltd.

Boom, R., Sol, C., Beld, M., Weel, J., Goudsmit, J. and Wertheim-van Dillen, P. (1999) f alpha-casein to silica particles. Journal of Clinical Microbiology 37(3), 615-619.

Boom, R., Sol, C.J.A., Salimans, M.M.M., Jansen, C.L., Wertheimvandillen, P.M.E. and Vandernoordaa, J. (1990) Rapid and simple method for purification of nucleic-acids. Journal of Clinical Microbiology 28(3), 495-503.

Bosch, A. (1998) Human enteric viruses in the water environment: A minireview. International Microbiology 1(3), 191-196.

Brow, C.N., Li, X., Rička, J. and Johnson, W.P. (2005) Comparison of microsphere deposition in porous media versus simple shear systems. Colloids and Surfaces A: Physicochemical and Engineering Aspects 253(1–3), 125-136.

Chung, J.W., Foppen, J.W., Gerner, G., Krebs, R. and Lens, P.N.L. (2015) Removal of rotavirus and adenovirus from artificial ground water using hydrochar derived from sewage sludge. Journal of Applied Microbiology, n/a-n/a.

Chung, J.W., Foppen, J.W., Izquierdo, M. and Lens, P.N.L. (2014) Removal of Escherichia coli from saturated sand columns supplemented with hydrochar produced from maize. Journal of Environmental Quality 43(6), 2096-2103.

Chung, J.W., Foppen, J.W. and Lens, P.N.L. (2013) Development of low cost two-step reverse transcription-quantitative polymerase chain reaction assays for rotavirus detection in foul surface water drains. Food and environmental virology 5(2), 126-133.

Davis, J.A. (1982) Adsorption of natural dissolved organic matter at the oxide/water interface. Geochimica et Cosmochimica Acta 46(11), 2381-2393.

Diener, S., Semiyaga, S., Niwagaba, C.B., Muspratt, A.M., Gning, J.B., Mbéguéré, M., Ennin, J.E., Zurbrugg, C. and Strande, L. (2014) A value proposition: Resource recovery from faecal sludge—Can it be the driver for improved sanitation? Resources, Conservation and Recycling 88(0), 32-38.

Elimelech, M., Nagai, M., Ko, C.H. and Ryan, J.N. (2000) Relative insignificance of mineral grain zeta potential to colloid transport in geochemically heterogeneous porous media. Environmental Science & Technology 34(11), 2143-2148.

Elimelech, M. and Omelia, C.R. (1990) Kinetics of deposition of colloidal particles in porous-media. Environmental Science & Technology 24(10), 1528-1536.

Farthing, M. (1989) Gut viruses: a role in gastrointestinal disease?, Smith Kline & French, Ltd., Welwyn Garden City, Hertfordshire, United Kingdom.

Foppen, J.W., van Herwerden, M. and Schijven, J. (2007) Transport of Escherichia coli in saturated porous media: Dual mode deposition and intra-population heterogeneity. Water Research 41(8), 1743-1753.

Friefeld, B.R., Lichy, J.H., Field, J., Gronostajski, R.M., Guggenheimer, R.A., Krevolin, M.D., Nagata, K., Hurwitz, J. and Horwitz, M.S. (1984) The in vitro replication of adenovirus DNA. Current topics in microbiology and immunology 110, 221-255.

Funke, A. and Ziegler, F. (2010) Hydrothermal carbonization of biomass: A summary and discussion of chemical mechanisms for process engineering. Biofuels Bioproducts & Biorefining-Biofpr 4(2), 160-177.

Heilmann, S.M., Molde, J.S., Timler, J.G., Wood, B.M., Mikula, A.L., Vozhdayev, G.V., Colosky, E.C., Spokas, K.A. and Valentas, K.J. (2014) Phosphorus Reclamation through Hydrothermal Carbonization of Animal Manures. Environmental Science & Technology 48(17), 10323-10329.

Hernroth, B.E., Conden-Hansson, A.C., Rehnstam-Holm, A.S., Girones, R. and Allard, A.K. (2002) Environmental factors influencing human viral pathogens and their potential indicator organisms in the blue mussel, Mytilus edulis: the first Scandinavian report. Applied and Environmental Microbiology 68(9), 4523-4533.

Hutton, G., Haller, L. and Bartram, J. (2007) Global cost-benefit analysis of water supply and sanitation interventions. Journal of Water and Health 5(4), 481-502.

Israelachvili, J. and Wennerstrom, H. (1996) Role of hydration and water structure in biological and colloidal interactions. Nature 379(6562), 219-225.

Jeong, S.-B., Yang, Y.-C., Chae, Y.-B. and Kim, B.-G. (2009) Characteristics of the Treated Ground Calcium Carbonate Powder with Stearic Acid Using the Dry Process Coating System. Materials Transactions 50(2), 409-414.

John, D.E. and Rose, J.B. (2005) Review of factors affecting microbial survival in groundwater. Environmental Science & Technology 39(19), 7345-7356.

Johnson, W.P., Li, X. and Assemi, S. (2007) Deposition and re-entrainment dynamics of microbes and non-biological colloids during non-perturbed transport in porous media in the presence of an energy barrier to deposition. Advances in Water Resources 30(6-7), 1432-1454.

Katukiza, A.Y., Ronteltap, M., Niwagaba, C.B., Foppen, J.W.A., Kansiime, F. and Lens, P.N.L. (2012) Sustainable sanitation technology options for urban slums. Biotechnology Advances 30(5), 964-978.

Katukiza, A.Y., Temanu, H., Chung, J.W., Foppen, J.W.A. and Lens, P.N.L. (2013) Genomic copy concentrations of selected waterborne viruses in a slum environment in Kampala, Uganda. Journal of Water and Health 11(2), 358-370.

Koné, D. (2010) Making urban excreta and wastewater management contribute to cities' economic development: A paradigm shift. Water Policy 12(4), 602-610.

Li, X.Q., Zhang, P.F., Lin, C.L. and Johnson, W.P. (2005) Role of hydrodynamic drag on microsphere deposition and re-entrainment in porous media under unfavorable conditions. Environmental Science & Technology 39(11), 4012-4020.

Lodder, W.J., van den Berg, H., Rutjes, S.A., Bouwknegt, M., Schijven, J.F. and Husman, A.M.D. (2013) Reduction of bacteriophage MS2 by filtration and irradiation determined by culture and quantitative real-time RT-PCR. Journal of Water and Health 11(2), 256-266.

Matthess, G., Bedbur, E., Gundermann, K.O., Loof, M. and Peters, D. (1991) Investigation on filtration mechanisms of bacteria and organic particles in porous-media .1. Background and methods. Zentralblatt Fur Hygiene Und Umweltmedizin 191(1), 53-97.

Michen, B. and Graule, T. (2010) Isoelectric points of viruses. Journal of Applied Microbiology 109(2), 388-397.

Miles, L. and Kapos, V. (2008) Reducing greenhouse gas emissions from deforestation and forest degradation: Global land-use implications. Science 320(5882), 1454-1455.

Minani, J.M.V., Foppen, J.W. and Lens, P.N.L. (2014) Sorption of cadmium in columns of sand-supported hydrothermally carbonized particles. Water science and technology : a journal of the International Association on Water Pollution Research 69(12), 2504-2509.

Mumme, J., Eckervogt, L., Pielert, J., Diakite, M., Rupp, F. and Kern, J. (2011) Hydrothermal carbonization of anaerobically digested maize silage. Bioresource Technology 102(19), 9255-9260.

Pang, X.L.L., Lee, B., Boroumand, N., Leblanc, B., Preiksaitis, J.K. and Ip, C.C.Y. (2004) Increased detection of rotavirus using a real time reverse transcription-polymerase chain reaction (RT-PCR) assay in stool specimens from children with diarrhea. Journal of Medical Virology 72(3), 496-501.

Parashar, U.D., Bresee, J.S., Gentsch, J.R. and Glass, R.I. (1998) Rotavirus. Emerging Infectious Diseases 4(4), 561-570.

Redman, J.A., Walker, S.L. and Elimelech, M. (2004) Bacterial adhesion and transport in porous media: Role of the secondary energy minimum. Environmental Science & Technology 38(6), 1777-1785.

Schijven, J.F. and Hassanizadeh, S.M. (2000) Removal of viruses by soil passage: Overview of modeling, processes, and parameters. Critical Reviews in Environmental Science and Technology 30(1), 49-127.

Sobsey, M.D., Stauber, C.E., Casanova, L.M., Brown, J.M. and Elliott, M.A. (2008) Point of use household drinking water filtration: A practical, effective solution for providing sustained access to safe drinking water in the developing world. Environmental Science & Technology 42(12), 4261-4267.

Stewart, P.L., Fuller, S.D. and Burnett, R.M. (1993) Difference imaging of adenovirus - bridging the resolution gap between x-ray crystallography and electron-microscopy. Embo Journal 12(7), 2589-2599.

Sun, K., Ro, K., Guo, M., Novak, J., Mashayekhi, H. and Xing, B. (2011) Sorption of bisphenol A, 17α-ethinyl estradiol and phenanthrene on thermally and hydrothermally produced biochars. Bioresource Technology 102(10), 5757-5763.

Tipping, E. and Cooke, D. (1982) The effects of adsorbed humic substances on the surface charge of goethite (α-FeOOH) in freshwaters. Geochimica et Cosmochimica Acta 46(1), 75-80.

Tong, M. and Johnson, W.P. (2006) Excess Colloid Retention in Porous Media as a Function of Colloid Size, Fluid Velocity, and Grain Angularity. Environmental Science & Technology 40(24), 7725-7731.

WHO and UNICEF (2014) Progress on drinking water and sanitation: 2014 update, WHO, Geneva, Switzerland / UNICEF, New York, USA. ISBN: 978-92-4-150724-0.

Yezli, S. and Otter, J.A. (2011) Minimum Infective Dose of the Major Human Respiratory and Enteric Viruses Transmitted Through Food and the Environment. Food and Environmental Virology 3(1), 1-30.

Chapter 8: General discussion and conclusions

8.1 PCR-based methods for pathogen removal assessments

The virus concentration can be measured by using either culture-based methods or molecular detection methods. Culture-based methods are sensitive at infectivity of viruses, which determines the microbial risk. However, it is difficult to differentiate a single virion from an aggregate, and this might result in an underestimation of the virus concentration (Teunis et al., 2005). Also, culturing may take days to weeks to yield the results (Pecson et al., 2009). In contrast, molecular detection methods are rapid and able to quantify viruses in aggregates within a day (Husman et al., 2009). The main limitation of the molecular detection methods for the virus removal technology is their inability of distinguishing infectious and non-infectious virions. Due to the fact that nucleic acid can be very persistent in the environment, residual target templates from inactivated virons can also give positive signals, leading to an overestimation of their number (Josephson et al., 1993; Masters et al., 1994).

In this PhD thesis, we developed low cost home-made (RT)-qPCR assays with competitive sensitivity to commercial kits (Chapter 3 and 6), and these assays were employed in the virus removal experiments (Chapter 6 and 7). The results from these chapters must be interpreted as a worst-case scenario, assuming that all virions in the samples were infectious. Lodder et al. (2013) reported significant differences between the standard plaque forming unit (PFU) method and a RT-qPCR assay; in the tail of the breakthrough curve (BTC) the virus concentration obtained from RT-qPCR was approximately 2.5 times higher than those detected by the PFU method. Likewise, the effluent samples in this research might have contained virions which lost their infectivity in column experiments, but still contained nucleic acid sequences targeted by PCR primers. However, the significance of false-positive results in this research was difficult to assess, because Lodder et al. (2013) employed much longer filtration times (a few days) than those applied in this research (within 5 h).

A virus loses its infectivity because of disruption of surface proteins and/or degradation of nucleic acid (Gerba, 1984). The inactivation may occur when viruses are suspended in aqueous solutions or associated with surfaces. Figure 8.1 presents potential mechanisms of virus inactivation in aqueous solutions (e.g. groundwater) (Harvey and Ryan, 2004). Since viruses are composed of labile proteins, natural die-off of viruses occurs in most environmental conditions. Adverse conditions, e.g. high temperature, pH and microbial activity, can facilitate inactivation of viruses through degradation of the viral genome and/or receptors (host-recognition site)(Schijven and Hassanizadeh, 2000; Harvey and Ryan, 2004).

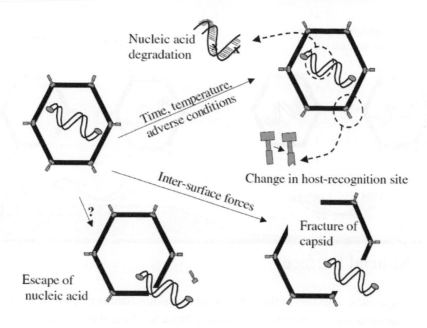

Figure 8.1: Schematic depiction of different mechanisms by which viruses may lose their infectivity while suspended in solution (modified from Harvey and Ryan, 2004).

The effect of association of a virus particle with a solid surface on virus inactivation depends on the type of the solid surface. It was reported that attachment to geological media with high organic content and clay-sized particles delayed the virus inactivation (Stagg et al., 1977; Straub et al., 1992). Also, attachment on solid surfaces provided resistance to disinfectants (EPA, 1999). Schijven (1999) reported that PRD1 inactivation was slightly slower in dune sand filtration than when the virions were suspended in the solution. In contrast, the attachment on solid media which provides strong binding forces, e.g. metal oxides, accelerated virus inactivation (Figure 8.2). It was reported that strong electrostatic attraction between viruses and surfaces of the opposite charge induced a disruption of the virus structure (Murray and Laband, 1979). In contrast, the binding of viruses attached on negatively charged surfaces was much weaker. Therefore, the virus inactivation was either not observed or retarded (Loveland et al., 1996; Harvey and Ryan, 2004). In order to provide more information on the infectivity status of viruses detected by PCR-based techniques, pretreatment of samples with several chemicals such as Protenase K, RNAse, Propidium Monoazide and Ethidium Monoazide were suggested (Nuanualsuwan and Cliver, 2002; Nocker et al., 2007; Baert et al., 2008; Seinige et al., 2014).

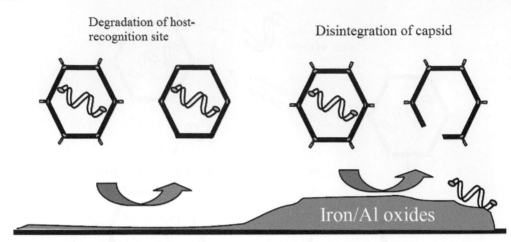

Figure 8.2: Schematic depiction of the different potential mechanisms by which viruses may lose their infectivity in an association with surfaces (modified from Harvey and Ryan, 2004)

Interesting future research topics include the infectivity assessments on viruses either attached on the hydrochar surface or contained in the effluent. Comparison between standard culture-based methods and PCR-based techniques (with or without pretreatment) can provide important information for designing water treatment systems using hydrochar adsorbents.

8.2   Hydrochar application in water treatment

Hydrochar has been considered as an energy-efficient method converting various organic wastes into valuable carbonaceous materials. Hydrochar has a wide range of applications including: carbon sequestration, soil amendment, nano-structured materials, catalysts, energy production and storage (Libra et al., 2011). Also, recent studies suggested its application as an adsorbent in water and wastewater treatment for removal of various abiotic contaminants such as heavy metals, radioactive agents, organic dye, phosphorus and pharmaceutics. Table 8.1 summarizes the hydrochar production parameters and decontamination capacity of selected hydrochar adsorbents as well as the results from this dissertation investigating the pathogen removal.

The contaminant removal capacity of adsorbents is proportional to an accessible surface area and surface functionality favourable for target contaminants. Compared to the conventional activated carbon products, hydrochars have a relatively limited surface area that is unfavorable for the adsorptive removal. However, the richness in oxygenated surface functional groups of the hydrochar compensate this shortcoming (Titirici et al., 2012). It was reported that high temperature and long residence time in the hydrochar production process had a negative impact on the surface porosity. The porous structure was cracked down and pores were partially blocked by repolymerization / recondensation of water-soluble compounds (Liu et al., 2010). The observation of Parshetti et al. (2014) agreed with this phenomenon: two types of hydrochars were produced from urban food waste at different temperatures (250 or 350 'C) as adsorbents for textile dye removal from wastewater. The results showed that the hydrochar produced at low temperature had a higher specific surface area and more diverse surface functional groups, leading to a higher adsorption capacity than the one produced at the higher temperature. However, our observation from the faecal waste-derived hydrochar produced at two distinctive reaction severities (temperature and residence time, Chapter 7) did not show differences in virus removal efficiency (Table 7.2). It is difficult to draw a general conclusion about the relation between the HTC reaction conditions and virus adsorption capacities. The performance of hydrochar adsorbents could be dependent on the nature of the feedstock and the target contaminants. Therefore, the optimization of HTC parameters is strongly recommended before large scale implementation. Also, a larger energy demand for higher reaction temperatures and longer residence times must be taken into consideration.

Relatively simple post treatments were applied to improve the performance of hydrochar adsorbents. Treatment with KOH or NaOH increased the adsorptive capacity of hydrochar for heavy metal and *E. coli* removal (Regmi et al., 2012; Islam et al., 2015; Chapter 4 and 5). It was speculated that strong alkali solutions removed tar-like substances from the hydrochar surface, leading to an increase in surface area and active adsorption sites. Also, magnetization of hydrochar using a $FeCl_3$ solution enhanced the recovery of saturated adsorbents from aqueous solutions, which can be beneficial for operating a water treatment unit (Zhu et al., 2014a).

Table 8.1: Hydrochar used in water treatment as an adsorbent

Feedstock	HTC conditions				Contaminants / adsorption capacity (mg / g) or pathogen / removal efficiency (%)	Description	
	Temp. (°C)	Duration (hour)	Pressure (bar)	Dry matter (%)		Remarks	Reference
Seed shell (*Prosopis Africana*)	200	0.33	18	6	Pb²⁺ / 45.3 Cd²⁺ / 38.3	High frequency microwave was used for HTC. Hydrochar showed better removal than biochar produced from the same feedstock.	Elaigwu et al., 2014
Clay (Palygorskite) + Cellulose	210	12	ᵃn/a	9	Phenol / 0.5	The clay minerals were coated with carbon nano-particles through co-carbonization.	Wu et al., 2014
					Cd²⁺ / 34	The adsorption capacity of hydrochar increased by activation with KOH	Regmi et al., 2012
					Cu²⁺ / 31		
Switchgrass	300	0.5	n/a	88	U(VI) / 2.1	Richness in active functional groups facilitated the adsorptive removal of Uranium	Kumar et al., 2011
Food waste	250	0.33	n/a	25	Acridine Orange / 95	The hydrochar produced at lower temperature showed higher specific surface are and better larger adsorption capacity	Parshetti et al., 2014
Peanut hull	300	5	69	13	Pb²⁺ / 22.8	The adsorption capacity of hydrochar increased by activation with H₂O₂	Xue et al., 2012
Willow (*salix psammophila*)	300	n/a	n/a	8	Tetracycline / 17	Hydrochar was magnetized by γ-Fe2O3 treatment for the ease of adsorbent separation	Zhu et al., 2014b

ᵃ n/a - not available

Table 8.1 continues

Feedstock	HTC conditions				Contaminants / adsorption capacity (mg / g) or pathogen / removal efficiency (%)	Description	
	Temp. (°C)	Duration (hour)	Pressure (bar)	Dry matter (%)		Remarks	Reference
Palm date seed	200	5	n/a	5	Methylen Blue dye / 621.1	The adsorption capacity of hydrochar increased by activation with NaOH	Islam et al., 2015
Maize	n/a	n/a	n/a	n/a	Cd^{2+} / 0.1	Despite relatively low adsorption capacity, removal efficiencies in column experiment high (~90%) during 20 pore volumes of flushing.	Minani et al., 2014
					Escherichia coli / 93	The adsorption capacity of hydrochar increased by activation with KOH	Chapter 4
					Cu^{2+} / 30 PO_4^{3-} / 3.7	The adsorption capacity of hydrochar increased by activation with KOH	Spataru, 2014
Sewage sludge	210	5	22	n/a	Escherichia coli / 90	The adsorption capacity of hydrochar increased by activation with KOH	Chapter 5
					Rotavirus / 99 Adenovirus / 99	HTC was carried out for safe disposal of stabilized sewage sludge from wastewater treatment plant	Chapter 6
Swine faeces	230 180	7 2	37 12	15.8	Rotavirus / 99 Adenovirus / 99	Hydrochars produced from different conditions showed comparable removal efficiency in column experiments	Chapter 7

[a] n/a - not available

These additional treatments may, however, not be desirable due to extra costs and environmental burdens from the by-products. Instead, hydrothermal co-carbonization can be suggested as an alternative for chemical activation methods improving the quality of hydrochar adsorbents. Wu et al. (2014) reported that co-carbonization of clay minerals and cellulose produced a capable adsorbent for phenol removal from aqueous solutions. During the repolymerization / recondensation process in HTC, clay minerals were coated with nano carbon which provided organophilic functional groups, improving the phenol removal performance of the adsorbent. Also, co-carbonization could prevent the release of undesirable products from hydrochar. Addition of rice husk in HTC of sewage sludge decreased the release of heavy metals in leaching experiments (Shi et al., 2013).

Bacterial adhesion / attachment on the carbonaceous surfaces facilitated biofilm formation and growth (Tobin et al., 1981; Rollinger and Dott, 1987). Due to the risk of multiplication and the likelihood of an episodic release of the accumulated bacteria, the fate of pathogens adhered on the adsorbent surface should be carefully considered in water treatment systems (Percival and Walker, 1999). The fate of adhered bacteria is affected by the type of bacteria and the electrical charge of the solid surface. In general, positively charged surfaces are reported to have greater adhesion of bacteria, while also inhibiting of biofilm formation due to inactivation of adhered bacteria (Busscher et al., 2006; van der Mei et al., 2008a; van der Mei et al., 2008b). The positive surface charge of an activated carbon surface induced cell membrane disruption, and thus results in better inactivation of adhered bacteria compared to negatively charged carbonaceous surfaces (van der Mei et al., 2008a). In addition to the insignificant bactericidal impact, negatively charged surfaces might promote the detachment of the biofilm from the surfaces due to the electrostatically repulsive conditions (Terada et al., 2012). Chapter 4 and 5 of this dissertation suggested the use of hydrochar as an adsorbent for *E. coli* removal. However, the survival of *E. coli* attached on the hydrochar adsorbent was not studied in this PhD research. It could be speculated that the inactivation of *E. coli* would be retarded due to net-negative surface charge of hydrochar adsorbents (Figure 4.2 and 5.1). The same concern applies to viruses attached on the hydrochar adsorbents. Though viruses do not multiply outside the host, it can still remain infectious when attached on the hydrochar, posing microbial risks either in water consumption or in disposal of virus-saturated adsorbents. Therefore, the fate of pathogens adhered on the hydrochar adsorbents is an important research topic for water quality surveillance. In the same context, safe disposal methods for pathogen-saturated hydrochar need to be developed. To obtain sustainable water safety the concept of Hazard Analysis and Critical Control Points (HACCP) can be recommended. Havelaar (1994) described the application of HACCP to drinking-water supply. It considered pollution of raw water sources, recontamination during storage and distribution for treated water and growth of pathogen in raw and treated waters. In addition to this framework, important contents for the HACCP in water treatment systems using hydrochar adsorbent include: separation of the virus containing spent hydrochar from the water treatment unit, safe delivery and / or storage and final disposal methods.

The low-cost prospect of the HTC technology and the less energy dependent conversion of biowastes into capable adsorbents can be an attractive feature in developing countries, where the implementation of conventional treatment technologies is difficult. In those areas, Point-Of-Use (POU) water treatment technologies (e.g. biosand filtration, ceramic filtration, solar disinfection and chlorination-coagulation) have been extensively implemented to ensure biological drinking-water safety (Walker et al., 2004; Sobsey et al., 2008; Bielefeldt et al., 2009). However, they have relatively limited efficiencies for removal of pathogenic viruses (Sobsey et al., 2008). Also, coagulation-chlorination techniques can induce adverse health impacts. The carcinogenic disinfection by-products can be generated by chlorination of water with high organic matter content (McDonald and Komulainen, 2005; Melnick et al., 2007). The residual aluminium concentration in drinking water treated by aluminium-salt based coagulants has been put in controversial debate for its link to the Alzheimer's disease (Flaten, 2001). The multi-barrier approach combining these conventional POU technologies and hydrochar adsorbents has a potential for mitigation of these shortcomings. The results from Chapter 6 and 7 describe efficient virus removal of hydrochar derived from sewage sludge and faecal sludge (Table 6.2 and 7.2). Also, the hydrochar adsorbents used for pathogen removal in Chapter 4, 5 and 6 showed an efficient removal of heavy metals (Minani et al., 2014; Spataru, 2014). In addition to coagulation / chlorination techniques, successive treatment with hydrochar as a post-treatment option can provide the removal of residual aluminium compounds as well as remaining pathogens. Some interesting topics for further research include the simultaneous removal of both biotic and abiotic contaminants of interest. This can be an ideal option for specific cases of water contamination in countries like Bangladesh, where the ground water is contaminated by both faecal microorganism and arsenic (Ferguson et al., 2012).

8.3   Hydrothermal carbonization for sewage sludge treatment

Sewage sludge produced from wastewater treatment plants has been traditionally used as a fertilizer or disposed of in landfills. Despite of its capability of supplying nutrients (e.g. phosphorus and nitrogen) in agricultural fields, sewage sludge often contains organic substances, pathogenic microorganisms and heavy metals (Sterritt et al., 1980). Escala et al. (2012) suggested HTC as a more energy-efficient measure for the safe disposal of sewage sludge with a possible application of the resulting hydrochar for as a fuel for energy production. In a broader perspective, "Sewchar" concept was suggested as a sustainable treatment and recycling option for sewage sludge and human waste (Breulmann et al., 2015). Figure 8.3 depicts the material flow and potential benefits of the Sewchar.

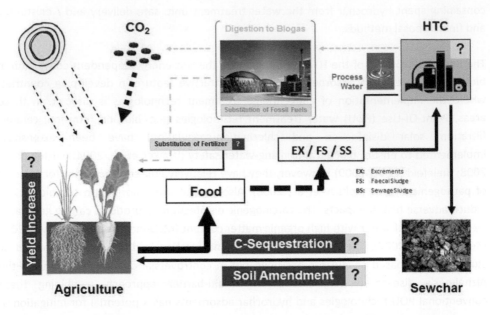

Figure 8.3: Schematic diagram of "Sewchar" concept (www.ufz.de/index.php?en=31930)
The implementation of the sewchar concept would efficiently mitigate the global climate change by: (i) carbon sequestration in soil (decreasing greenhouse gas emission), (ii) increasing biomass productivity, (iii) the saving or substitution of fossil fuels. Also, the potential benefits may include improvement of resource recovery in wastewater treatment systems, degradation of organic pollutants and pathogenic microorganisms (Breulmann et al., 2015). In addition, the results from Chapter 5, 6 and 7 provide evidence for possible use of resulting hydrochar as a capable adsorbent. Conceptually, the generation of sewchar and its subsequent utilization in water and/or wastewater treatment would develop more closed material cycle in wastewater treatment systems.

8.4 Hydrothermal carbonization for sanitation and water treatment

Affordable faecal waste, wastewater and water treatment systems are essential needs for every community. Untreated human faeces in less developed communities are considered as the main source of pathogens in surface water and groundwater (Katukiza et al., 2013). As a new solution for sanitation and water problems of places in need, this research suggested HTC of faecal waste and the application of resulting hydrochar in water treatment (Chapter 7). Figure 8.4 depicts a virus circulation route in less developed communities and the potential of faecal waste-HTC technology for solving water-sanitation problems.

Faecal waste, the most important source of human pathogens, will be totally sanitized during the HTC process with elevated temperature (~200 °C) and pressure (~20 bar), and the resulting hydrochar can be used as a virus adsorbent in water treatment (Chapter 7). Also, the applicability of wet feedstock makes the faeces - HTC technique an attractive alternative to conventional faecal sludge drying beds when they malfunction in rainy seasons.

One of the most important challenges of implementing the faecal waste-HTC in less developed communities can be a need for an affordable HTC reactor. Many technologies developed for less developed countries have failed to be customized in the places in need due to top-down and giveaway approach of technology development and transplantation (Polak and Warwick, 2013). Regarding the fact that the most vulnerable communities facing water and sanitation problems are also limited by technical and financial status, new technology must be developed based on locally available materials and human resources.

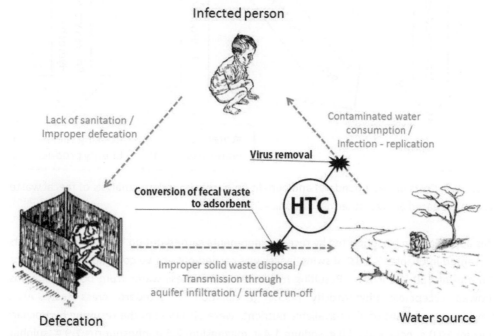

Figure 8.4: Virus circulation in less developed areas with poor faecal waste management and the potential of HTC technology to act as contaminant barrier.

The HTC-based faecal waste treatment can be a viable option when the cost needed for the HTC treatment is lower than current measures (if present), or the benefits from the HTC products exceed its cost (Koné, 2010). Figure 8.5 depicts the cost - benefit structure of faecal sludge treatment through HTC. The products from faecal waste-HTC can be utilized as agricultural supplement (Berge et al., 2013), energy source (Mumme et al., 2011) or adsorbent in water and / or wastewater treatment for pathogenic microorganisms (Chapter 4, 5, 6 and 7; Minani et al., 2014; Spataru 2014).

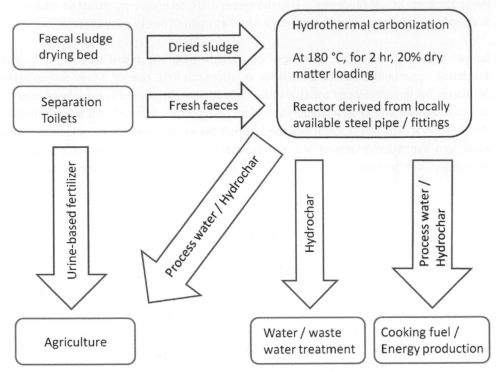

Figure 8.5: Costs (in red boundary) and benefit (in purple boundary) analysis of faecal waste treatment based on hydrothermal carbonization

This approach can be supported by recent experiments carried out on hydrochar adsorbents presented in Chapter 7. HTC of swine faeces was carried out at two conditions: at 220°C for 7 h and at 180 °C for 2 h. Resulting hydrochar and process water from both conditions showed acceptable phytotoxicity (>95% germination in standard cress test). Also, considerable amount of plant-available nutrients were observed in the resulting hydrochar: calcium 50.0 g, potassium 6.0 g, sodium 1.4 g, magnesium 2.5 g, phosphorus 7.9 g, sulphur 9.8 g / kg hydrochar (unpublished data). Since the characteristics of hydrochar and process water are determined by HTC conditions (temperature, pressure and water content) and feed stock composition, the quality of the products can be improved by optimization of the HTC process (Libra et al., 2011). Important contents for the evaluation of the HTC products are summarized in Table 8.3.

Table 8.3: Potential use of faecal waste HTC products and relevant investigations

Potential usage	HTC component	
	Hydrochar	Process water
Agricultural supplements	plant-available nutrients, generation of hazardous by-products, phytotoxicity and effect on local crop production	
Energy source	calorific value, generation of hazardous exhaust gas from combustion	methane production capacity in biogas plant
Water / Wastewater treatment	adsorptive removal of contaminants, leaching of hazardous materials, safe disposal methods of hydrochar saturated with contaminants	Not applicable

8.4 Conclusions

The following main conclusions can be made from the research carried out in this dissertation:

i)   Home-made RT-qPCR assays to detect rotaviruses showed comparable efficiency and sensitivity to a selected commercial kit in both laboratory samples and environmental samples. Homemade assays were cheaper (ca. 10 times) than the selected commercial kit, but required longer reaction times and more manual operations. The home-made assays can be useful in those laboratories where the costs for the reagents are hampering the laboratory in its functioning more than the costs for human labour.

ii)   The *E. coli* removal efficiency of sand columns supplemented with alkali-activated hydrochar derived from agricultural waste of maize or sewage sludge showed an improved *E. coli* removal efficiency of ~90 %. Apparently, the activation with KOH removed the water-soluble organic substances deposited on the surface of the hydrochar, resulting in development of macroporous surfaces or increased hydrophobicity.

iii)   Hydrochar derived from fresh swine waste or stabilized sewage sludge from a wastewater treatment plant showed a remarkably high pathogenic rotavirus and adenovirus removal efficiency. Sand columns of 10 cm bed height supplemented with the hydrochar (1.5 % *w/*w) could successfully remove more than 99 % of both viruses.

iv)   Despite similar size and shape of rotavirus and adenovirus, they showed a different dependency on their retention in the secondary energy minimum. In deionized water flushing experiments, more release was observed for rotavirus than adenovirus throughout the whole research. Apparently, the contribution of the secondary energy minimum in the virus retention was more determined by the surface characteristics of the virions than their size and shape.

8.5 References

Baert, L., Wobus, C.E., Van Coillie, E., Thackray, L.B., Debevere, J. and Uyttendaele, M. (2008) Detection of Murine Norovirus 1 by Using Plaque Assay, Transfection Assay, and Real-Time Reverse Transcription-PCR before and after Heat Exposure. Applied and Environmental Microbiology 74(2), 543-546.

Berge, N.D., Kammann, C., Ro, K. and Libra, J. (2013) Sustainable Carbon Materials from Hydrothermal Processes, pp. 295-340, John Wiley & Sons, Ltd.

Bielefeldt, A.R., Kowalski, K. and Summers, R.S. (2009) Bacterial treatment effectiveness of point-of-use ceramic water filters. Water Research 43(14), 3559-3565.

Breulmann, M., Afferden, M.v., Müller, R.A. and Fühner, C. (2015) The Sewchar Concept: An Innovative Tool for a Sustainable Water - Waste - Soil Nexus of Sanitation System, Dresden, Germany.

Busscher, H.J., Dijkstra, R.J.B., Engels, E., Langworthy, D.E., Collias, D.I., Bjorkquist, D.W., Mitchell, M.D. and van der Mei, H.C. (2006) Removal of two waterborne pathogenic bacterial strains by activated carbon particles prior to and after charge modification. Environmental Science & Technology 40(21), 6799-6804.

Dewettinck, T., Van Houtte, E., Geenens, D., Van Hege, K. and Verstraete, W. (2001) HACCP (Hazard Analysis and Critical Control Points) to guarantee safe water reuse and drinking water production - a case study. Water Science and Technology 43(12), 31-38.

Elaigwu, S.E., Rocher, V., Kyriakou, G. and Greenway, G.M. (2014) Removal of Pb2+ and Cd2+ from aqueous solution using chars from pyrolysis and microwave-assisted hydrothermal carbonization of Prosopis africana shell. Journal of Industrial and Engineering Chemistry 20(5), 3467-3473.

EPA (1999) EPA Guidance Manual Turbidity Provisions.

Escala, M., Zumbuhl, T., Koller, C., Junge, R. and Krebs, R. (2013) Hydrothermal Carbonization as an Energy-Efficient Alternative to Established Drying Technologies for Sewage Sludge: A Feasibility

Ferguson, A.S., Layton, A.C., Mailloux, B.J., Culligan, P.J., Williams, D.E., Smartt, A.E., Sayler, G.S., Feighery, J., McKay, L.D., Knappett, P.S.K., Alexandrova, E., Arbit, T., Emch, M., Escamilla, V., Ahmed, K.M., Alam, M.J., Streatfield, P.K., Yunus, M. and van Geen, A. (2012) Comparison of fecal indicators with pathogenic bacteria and rotavirus in groundwater. Science of the Total Environment 431(0), 314-322.

Flaten, T.P. (2001) Aluminium as a risk factor in Alzheimer's disease, with emphasis on drinking water. Brain Research Bulletin 55(2), 187-196.

Gerba, C.P. (1984) Advances in Applied Microbiology. Allen, I.L. (ed), pp. 133-168, Academic Press.

Harvey, R.W. and Ryan, J.N. (2004) Use of PRDI bacteriophage in groundwater viral transport, inactivation, and attachment studies. Fems Microbiology Ecology 49(1), 3-16.

Havelaar, A.H. (1994) Application of HACCP to drinking water supply. Food Control 5(3), 145-152.

Husman, A., Lodder, W.J., Rutjes, S.A., Schijven, J.F. and Teunis, P.F.M. (2009) Long-Term Inactivation Study of Three Enteroviruses in Artificial Surface and Groundwaters, Using PCR and Cell Culture. Applied and Environmental Microbiology 75(4), 1050-1057.

Islam, M.A., Tan, I.A.W., Benhouria, A., Asif, M. and Hameed, B.H. (2015) Mesoporous and adsorptive properties of palm date seed activated carbon prepared via sequential hydrothermal carbonization and sodium hydroxide activation. Chemical Engineering Journal 270, 187-195.

Josephson, K.L., Gerba, C.P. and Pepper, I.L. (1993) Polymerase chain-reaction detection of nonviable bacterial pathogens. Applied and Environmental Microbiology 59(10), 3513-3515.

Katukiza, A.Y., Temanu, H., Chung, J.W., Foppen, J.W.A. and Lens, P.N.L. (2013) Genomic copy concentrations of selected waterborne viruses in a slum environment in Kampala, Uganda. Journal of Water and Health 11(2), 358-370.

Koné, D. (2010) Making urban excreta and wastewater management contribute to cities' economic development: A paradigm shift. Water Policy 12(4), 602-610.

Kumar, S., Loganathan, V.A., Gupta, R.B. and Barnett, M.O. (2011) An Assessment of U(VI) removal from groundwater using biochar produced from hydrothermal carbonization. Journal of Environmental Management 92(10), 2504-2512.

Libra, J.A., Ro, K.S., Kammann, C., Funke, A., Berge, N.D., Neubauer, Y., Titirici, M.M., Führer, C., Bens, O., Kern, J. and Emmerich, K.H. (2011) Hydrothermal carbonization of biomass residuals: A comparative review of the chemistry, processes and applications of wet and dry pyrolysis. Biofuels 2(1), 71-106.

Liu, Z., Zhang, F.-S. and Wu, J. (2010) Characterization and application of chars produced from pinewood pyrolysis and hydrothermal treatment. Fuel 89(2), 510-514.

Lodder, W.J., van den Berg, H., Rutjes, S.A., Bouwknegt, M., Schijven, J.F. and Husman, A.M.D. (2013) Reduction of bacteriophage MS2 by filtration and irradiation determined by culture and quantitative real-time RT-PCR. Journal of Water and Health 11(2), 256-266.

Loveland, J.P., Ryan, J.N., Amy, G.L. and Harvey, R.W. (1996) The reversibility of virus attachment to mineral surfaces. Colloids and Surfaces A: Physicochemical and Engineering Aspects 107(0), 205-221.

Masters, C.I., Shallcross, J.A. and Mackey, B.M. (1994) Effect of stress treatments on the detection of listeria-monocytogenes and enterotoxigenic escherichia-coli by the polymerase chain-reaction. Journal of Applied Bacteriology 77(1), 73-79.

McDonald, T.A. and Komulainen, H. (2005) Carcinogenicity of the chlorination disinfection by-product MX. Journal of Environmental Science and Health Part C-Environmental Carcinogenesis & Ecotoxicology Reviews 23(2), 163-214.

Melnick, R.L., Nyska, A., Foster, P.M., Roycroft, J.H. and Kissling, G.E. (2007) Toxicity and carcinogenicity of the water disinfection byproduct, dibromoacetic acid, in rats and mice. Toxicology 230(2-3), 126-136.

Minani, J.M.V., Foppen, J.W. and Lens, P.N.L. (2014) Sorption of cadmium in columns of sand-supported hydrothermally carbonized particles. Water science and technology : a journal of the International Association on Water Pollution Research 69(12), 2504-2509.

Mumme, J., Eckervogt, L., Pielert, J., Diakite, M., Rupp, F. and Kern, J. (2011) Hydrothermal carbonization of anaerobically digested maize silage. Bioresource Technology 102(19), 9255-9260.

Murray, J.P. and Laband, S.J. (1979) Degradation of poliovirus by adsorption on inorganic surfaces. Appl Environ Microbiol 37(3), 480-486.

Nocker, A., Sossa, K.E. and Camper, A.K. (2007) Molecular monitoring of disinfection efficacy using propidium monoazide in combination with quantitative PCR. Journal of Microbiological Methods 70(2), 252-260.

Nuanualsuwan, S. and Cliver, D.O. (2002) Pretreatment to avoid positive RT-PCR results with inactivated viruses. Journal of Virological Methods 104(2), 217-225.

Parshetti, G.K., Chowdhury, S. and Balasubramanian, R. (2014) Hydrothermal conversion of urban food waste to chars for removal of textile dyes from contaminated waters. Bioresource Technology 161, 310-319.

Pecson, B.M., Martin, L.V. and Kohn, T. (2009) Quantitative PCR for Determining the Infectivity of Bacteriophage MS2 upon Inactivation by Heat, UV-B Radiation, and Singlet Oxygen: Advantages and Limitations of an Enzymatic Treatment To Reduce False-Positive Results. Applied and Environmental Microbiology 75(17), 5544-5554.

Percival, S.L. and Walker, J.T. (1999) Potable water and biofilms: a review of the public health implications. Biofouling 14(2), 99-115.

Polak, P. and Warwick, M. (2013) The Business Solution to Poverty: Designing Products and Services for Three Billion New Customers, Berrett-Koehler Publishers.

Regmi, P., Moscoso, J.L.G., Kumar, S., Cao, X.Y., Mao, J.D. and Schafran, G. (2012) Removal of copper and cadmium from aqueous solution using switchgrass biochar produced via hydrothermal carbonization process. Journal of Environmental Management 109, 61-69.

Rollinger, Y. and Dott, W. (1987) Survival of selected bacterial species in sterilized activated carbon filters and biological activated carbon filters. Applied and Environmental Microbiology 53(4), 777-781.

Schijven, J.F. and Hassanizadeh, S.M. (2000) Removal of viruses by soil passage: Overview of modeling, processes, and parameters. Critical Reviews in Environmental Science and Technology 30(1), 49-127.

Schljven, J.F., Hoogenboezem, W., Hassanizadeh, S.M. and Peters, J.H. (1999) Modeling removal of bacteriophages MS2 and PRD1 by dune recharge at Castricum, Netherlands. Water Resources Research 35(4), 1101-1111.

Seinige, D., Krischek, C., Klein, G. and Kehrenberg, C. (2014) Comparative Analysis and Limitations of Ethidium Monoazide and Propidium Monoazide Treatments for the Differentiation of Viable and Nonviable Campylobacter Cells. Applied and Environmental Microbiology 80(7), 2186-2192.

Shi, W.S., Liu, C.G., Shu, Y.J., Feng, C.P., Lei, Z.F. and Zhang, Z.Y. (2013) Synergistic effect of rice husk addition on hydrothermal treatment of sewage sludge: Fate and environmental risk of heavy metals. Bioresource Technology 149, 496-502.

Sobsey, M.D., Stauber, C.E., Casanova, L.M., Brown, J.M. and Elliott, M.A. (2008) Point of use household drinking water filtration: A practical, effective solution for providing sustained access to safe drinking water in the developing world. Environmental Science & Technology 42(12), 4261-4267.

Spataru, A. (2014) The use of hydrochar as a low cost adsorbent for heavy metal and phosphate removal from wastewater. Master's thesis, UNESCO-IHE, Delft, The Netherlands.

Stagg, C.H., Wallis, C. and Ward, C.H. (1977) Inactivation of clay-associated bacteriophage MS-2 by chlorine. Applied and Environmental Microbiology 33(2), 385-391.

Sterritt, R.M. and Lester, J.N. (1980) The value of sewage sludge to agriculture and effects of the agricultural use of sludges contaminated with toxic elements: A review. Science of The Total Environment 16(1), 55-90.

Straub, T.M., Pepper, I.L. and Gerba, C.P. (1992) PERSISTENCE OF VIRUSES IN DESERT SOILS AMENDED WITH ANAEROBICALLY DIGESTED SEWAGE-SLUDGE. Applied and Environmental Microbiology 58(2), 636-641.

Terada, A., Okuyama, K., Nishikawa, M., Tsuneda, S. and Hosomi, M. (2012) The effect of surface charge property on Escherichia coli initial adhesion and subsequent biofilm formation. Biotechnology and bioengineering 109(7), 1745-1754.

Teunis, P.F.M., Lodder, W.J., Heisterkamp, S.H. and Husman, A.M.D. (2005) Mixed plaques: Statistical evidence how plaque assays may underestimate virus concentrations. Water Research 39(17), 4240-4250.

Titirici, M.-M., White, R.J., Falco, C. and Sevilla, M. (2012) Black perspectives for a green future: hydrothermal carbons for environment protection and energy storage. Energy & Environmental Science 5(5), 6796-6822.

Tobin, R.S., Smith, D.K. and Lindsay, J.A. (1981) Effects of activated carbon and bacteriostatic filters on microbiological quality of drinking water. Applied and Environmental Microbiology 41(3), 646-651.

van der Mei, H.C., Atema-Smit, J., Jager, D., Langworthy, D.E., Collias, D.I., Mitchell, M.D. and Busscher, H.J. (2008a) Influence of adhesion to activated carbon particles on the viability of waterborne pathogenic bacteria under flow. Biotechnology and bioengineering 100(4), 810-813.

van der Mei, H.C., Rustema-Abbing, M., Langworthy, D.E., Collias, D.I., Mitchell, M.D., Bjorkquist, D.W. and Busscher, H.J. (2008b) Adhesion and viability of waterborne pathogens on p-DADMAC coatings. Biotechnology and bioengineering 99(1), 165-169.

Walker, D.C., Len, S.V. and Sheehan, B. (2004) Development and evaluation of a reflective solar disinfection pouch for treatment of drinking water. Applied and Environmental Microbiology 70(4), 2545-2550.

Wu, X.P., Gao, P., Zhang, X.L., Jin, G.P., Xu, Y.Q. and Wu, Y.C. (2014) Synthesis of clay/carbon adsorbent through hydrothermal carbonization of cellulose on palygorskite. Applied Clay Science 95, 60-66.

Xue, Y.W., Gao, B., Yao, Y., Inyang, M., Zhang, M., Zimmerman, A.R. and Ro, K.S. (2012) Hydrogen peroxide modification enhances the ability of biochar (hydrochar) produced from hydrothermal carbonization of peanut hull to remove aqueous heavy metals: Batch and column tests. Chemical Engineering Journal 200, 673-680.

Zhu, X., Liu, Y., Qian, F., Zhou, C., Zhang, S. and Chen, J. (2014a) Preparation of magnetic porous carbon from waste hydrochar by simultaneous activation and magnetization for tetracycline removal. Bioresource Technology 154(0), 209-214.

Zhu, X.D., Liu, Y.C., Qian, F., Zhou, C., Zhang, S.C. and Chem, J.M. (2014b) Porous carbon materials from waste hydrochar for tetracycline adsorption. Abstracts of Papers of the American Chemical Society 247.

Summary

This PhD study has evaluated hydrochars derived from biowastes as adsorbents for pathogen removal in water treatment. Pathogen removal experiments were conducted by carrying out breakthrough analysis using a simple sand filtration set-up. Glass columns packed by 10 cm sand bed supplemented with minor amount of hydrochar (1.5 %, *w/w*) were flushed with artificial ground water seeded with test microorganisms at an upward flow rate of 1 mL / min. Either back flushing or deionized water flushing was performed at pathogen retaining columns in order to investigate the pathogen removal mechanism of hydrochar-amended sand columns.

Two home-made two-step reverse transcription-quantitative polymerase chain reaction assays were developed in order to quantify rotavirus in the samples from virus removal experiments. Since the total cost of the assays was mainly determined by the cost of reverse transcriptase, two reverse transcriptases with the lowest consumer price were employed. The efficiencies of home-made assays were comparable to a selected reference commercial kit in analyzing both environmental and laboratory samples, while the total cost of home-made assays was 11 times less. Though home-made assays necessitate more manual operations and time, the low-cost aspect might be appealing in those settings where the expenditure for consumables inhibits laboratories in their functioning.

Hydrochars produced via hydrothermal carbonization of agricultural residue from maize or stabilized sewage sludge were evaluated for adsorptive removal of *Escherichia coli*. In order to improve the removal capacity, the hydrochars were activated by being suspended in 1M KOH solution (5 g hydrochar / L) for 1 h at room temperature. The activation improved the *Escherichia coli* removal efficiency from ~70 to ~90 %. In addition, successive detachment experiments carried out by performing back flushing or deionized water flushing into sand columns supplemented with maize-derived hydrochar indicated that the strength of *Escherichia coli* attachment increased by KOH activation. Also, *Escherichia coli* removal of sewage sludge-derived hydrochar was evaluated in larger column with 50 cm filter bed for 30 days of intermittent operation. 3 pairs of columns packed with either sand, raw hydrochar or activated hydrochar showed average removal efficiency of 36.5% ± 10.1 (n=60), 24.4% ± 10.5 (n=56) and 91.2% ± 7.5 (n=60), respectively. Idle time of filtration unit did not affect the *Escherichia coli* removal efficiency of hydrochar-amended columns. The results from material characterization attributed the enhancement in *Escherichia coli* removal induced from the KOH activation to development of macroporous surface with increased hydrophobicity and surface charge. It was apparent that the activation with the KOH solution removed tar-like substances from hydrochar surface, resulting in exposure of hydrophobic core and development of rough surface structure. Also, deposition of K^+ ion in hydrochar was observed, which might have increased the surface charge.

The removal of human pathogenic rotavirus and adenovirus was investigated using hydrochar derived from stabilized sewage sludge or swine faecal waste. Throughout virus removal experiments, rotavirus and adenovirus showed comparable removal. At 1 mL / min flow rate, raw hydrochar (without KOH activation) derived from either feedstock showed mean virus removal efficiency from 2 to 3 > log removal (99 - 100 %). Also flow rates of 2.5 and 5 mL / min were tested using faecal waste-derived hydrochar. The virus removal efficiency remained still high (2.1 log - 3 log) at the elevated flow rates. We speculated that the improvement in virus removal derived from hydrochar supplement is induced by provision of extra hydrophobic surfaces in sand column media. Regardless the type of packing material, successive deionized water flushing into virus-retaining columns released more rotavirus than adenovirus, indicating larger role of the secondary energy minimum in rotavirus retention. It was remarkable, because both types of viruses are similar in their shape and size. This observation provides evidence that virus transport-retention behaviour could be mainly determined by surface characteristics of virus rather than its size and shape. In this sense, the use of model virus needs to be carefully considered when performing water treatment or pathogen transport experiments.

Successful removal of pathogenic virus using faecal waste-derived hydrochar highlights the potential of hydrothermal carbonization technology in less developed regions where modern water-sanitation systems are not affordable. Faecal waste, one of the most important pathogen sources, can be completely sanitized in elevated temperature and pressure during hydrothermal carbonization process. Moreover, the resulting hydrochar can be utilized at water or wastewater treatment. Despite general low-cost aspect of hydrothermal carbonization such as less energy dependency than dry pyrolysis and utilization of waste as a feedstock, the need for high-pressure reactor might hamper the implementation of the technology. Development of localized low-cost reactor, evaluation of hydrochar in its use at agriculture and/or energy production and overall economic analysis are recommended.

Curriculum Vitae

Jae Wook Chung was born on 16th June 1980, in Seoul, Republic of Korea. He served at Republic of Korea Army from 2001 to 2003. In 2007, he obtained Bachelor of Science in Environmental Engineering from Kyung-Hee Univiersity (Suwon, Republic of Korea). He studied at Ghent University (Gent, Belgium) from 2008 to 2010, obtaining a Master of Science degree with distinction in Environmental Sanitation. His MSc research was carried out in Laboratory of Microbial Ecology and Technology (LabMET). In January 2011, he started Ph.D research at UNESCO-IHE Institute for Water  Education and Wageningen University. During this PhD research period he won the Green Talents Competition in 2013 organized by Federal Ministry of Education and Research, Germany (BMBF). His main research interest focuses on affordable water treatment and sanitation technology in less developed communities, such as slum area and rural villages.

List of scientific publications

Chung, J. W., J. W. Foppen & P. N. L. Lens, (2013) Development of low cost two-step reverse transcription-quantitative polymerase chain reaction assays for rotavirus detection in foul surface water drains. *Food and environmental virology* 5: 126-133.

Chung, J. W., J. W. Foppen, M. Izquierdo, and P. N. L. Lens, (2014) Removal of *Escherichia coli* from saturated sand columns supplemented with hydrochar produced from maize. *Journal of Environmental Quality* 43(6), 2096-2103.

Chung, J. W., J. W. Foppen, G. Gerner, R. Krebs. and P. N. L. Lens, (2015) Removal of rotavirus and adenovirus from artificial ground water using hydrochar derived from sewage sludge. Journal of Applied Microbiology 119(3), 876-884.

Chung, J. W., G. Gerner, J. W. Foppen, R. Krebs and P. N. L. Lens, (submitted) Removal of *Escherichia coli* using hydrochar derived from hydrothermal carbonization of sewage sludge.

Chung, J. W., J. W. Foppen and P. N. L. Lens, (submitted) Simultaneous removal of rotavirus and adenovirus using sand columns supplemented with hydrochar derived from swine faeces.

Katukiza, A. Y., H. Temanu, J. W. Chung, J. W. A. Foppen & P. N. L. Lens, (2013) Genomic copy concentrations of selected waterborne viruses in a slum environment in Kampala, Uganda. *Journal of Water and Health* 11: 358-370.

Netherlands Research School for the
Socio-Economic and Natural Sciences of the Environment

D I P L O M A

For specialised PhD training

The Netherlands Research School for the
Socio-Economic and Natural Sciences of the Environment
(SENSE) declares that

Jae Wook Chung

born on 16 June 1980 in Seoul, Republic of Korea

has successfully fulfilled all requirements of the
Educational Programme of SENSE.

Wageningen, 30 October 2015

the Chairman of the SENSE board

Prof. dr. Huub Rijnaarts

the SENSE Director of Education

Dr. Ad van Dommelen

The SENSE Research School has been accredited by the Royal Netherlands Academy of Arts and Sciences (KNAW)

K O N I N K L I J K E   N E D E R L A N D S E

The SENSE Research School declares that Mr Jae Chung has successfully fulfilled all requirements of the Educational PhD Programme of SENSE with a work load of 37 EC, including the following activities:

SENSE PhD Courses

o   Basic Statistics (2011)

o   Environmental Research in Context (2011)

o   Research in Context Activity: 'Participating in GreenTalents Competition (2013) and guest researcher at Helmholtz Institute, Leipzig, Germany' (2014)

Other PhD and Advanced MSc Courses

o   Nanotechnology for water and wastewater, UNESCO-IHE Delft (2011)

o   Biofilms for technology, UNESCO-IHE Delft (2012)

o   UNESCO-IHE PhD week, Delft (2011-2013)

External training at a foreign research institute

o   High pressure reactor operation for hydrothermal carbonization, German Biomass Research Center (DBFZ), Germany (2014)

o   Characterization / evaluation of hydrochar for soil amendments, Helmholtz Centre for Environmental Research (UFZ), Germany (2014)

Management and Didactic Skills Training

o   Supervision of internship student (2011)

o   Supervision of two MSc students with thesis entitled 'Removal of *Escherichia coli* in sand columns supplemented with hydrochar derived from sewage sludge' (2013-2014) and 'The use of hydrochar as a low cost adsorbent for heavy metal and phosphate removal from wastewater' (2014)

o   Co-organising Microbiology laboratory sessions, UNESCO-IHE, Delft (2013-2014)

Oral Presentations

o   *Adsorptive removal of viruses and Cd from aqueous solutions using hydrochar derived from sewage sludge.* 4th International Conference on Research Frontiers in Chalcogen Cycle Science & Technology (G16), 28-29 May 2015, Delft, The Netherlands

SENSE Coordinator PhD Education

Dr. ing. Monique Gulickx

Printed and bound by CPI Group (UK) Ltd, Croydon, CR0 4YY

21/10/2024

01777094-0008